Hirakawa · Rothe
Shoukas · Tyberg (Eds.)

Veins

Their Functional Role in the Circulation

With 78 Figures

Springer Japan KK

SENRI HIRAKAWA, M.D.
Professor Emeritus of Medicine, Gifu University School of Medicine, Gifu, 500 Japan, and Professor of Nutrition, Kobe Women's University, Kobe, 654, Japan

CARL F. ROTHE, PH.D.
Professor, Department of Physiology and Biophysics, Indiana University School of Medicine, Indianapolis, IN 46202-5120 USA

ARTIN A. SHOUKAS, PH.D.
Professor Biomedical Engineering, The Johns Hopkins University School of Medicine, Baltimore, MA 21205, USA

JOHN V. TYBERG, M.D., PH.D.
Professor of Medicine and Medical Physiology, and Heritage Medical Scientist, Faculty of Medicine, The University of Calgary, Calgary, Alberta, T2N 4N1 Canada

ISBN 978-4-431-68387-2   ISBN 978-4-431-68385-8 (eBook)
DOI 10.1007/978-4-431-68385-8

Printed on acid-free paper

© Springer Japan 1993

Originally published by Springer-Verlag Tokyo Berlin Heidelberg New York in 1993
Softcover reprint of the hardcover 1st edition 1993

Preface

The 8th International Congress of Biorheology was held at the Pacifico Yokohama, Japan, a brand new, versatile convention center for international meetings, from August 3 through 8, 1992. There were many plenary lectures and symposia, one of which was entitled "Mechanics of the Venous System." It was at this symposium that we, the editors of this monograph, each presented papers.

We then moved to Gifu, Japan, for the Gifu Workshop on Veins and Vascular Capacitance. This was held on August 9, 1992, with the Second Department of Medicine, Gifu University School of Medicine, serving as the host. Nine papers were presented in the oral sessions and there were five poster presentations. This monograph, which is intended to provide a bird's-eye view of recent trends in studies of the venous system, is an outgrowth of the Gifu Workshop. While it is not exactly the Proceedings of that workshop, materials in the monograph were developed from ideas presented there.

The monograph includes: (a) discussions of the interactions between the endothelium and the smooth muscle of the venous wall, (b) descriptions of coronary venous flow-velocity patterns measured by a novel method, (c) descriptions of the passive transport of macromolecules and fluids across single venular capillaries, (d) structure-function correlations in venous walls and valves, (e) venous capacitance changes in experimental heart failure, together with a discussion of the muscle pump, (f) discussions of the effects of atrial natriuretic peptide (ANP) on venous distensibility in healthy human subjects, (g) a clinical study of forearm venous stiffness in chronic heart failure, (h) an experimental study of the degree to which the baroreflex modifies the effects of vasodilators on systemic capacitance vessels, (i) a detailed discussion of the interplay among vascular resistance and blood flow and volume pertaining to the regulation of hepatic vascular capacitance, (j) description of a novel method of determining nitroglycerin- and catecholamine-induced changes in the capacitance of the human pulmonary venous system, (k) a new method for constructing the human pulmonary venous-return curve, (l) descriptions of human pulmonary venous flow-velocity measurements, and (m) application of plethysmographic techniques to the evaluation of venous varicosity.

We are grateful to Dr. N. Toda, Professor of Pharmacology, Shiga University of Medical Science, for contributing an invited paper to this monograph.

We wish to thank Springer-Verlag Tokyo, Inc. for its support in publishing this monograph.

November, 1993

S. HIRAKAWA
C.F. ROTHE
A.A. SHOUKAS
J.V. TYBERG

Table of Contents

List of Contributors

Responsiveness of Isolated Veins to Vasoactive Substances

Noboru Toda and Tomio Okamura[1]

Abstract. Although attention is currently directed to responsiveness of capacitance vessel to chemical, neural, and physical stimuli as an important factor regulating the circulating blood volume and venous return, information available so far is limited. This chapter includes data on mechanical responses of isolated veins from dogs and monkeys to endogenous vasoactive substances, such as acetylcholine, histamine, angiotensin II, prostaglandin I_2 and nitric oxide. It also compares responses of veins and arteries obtained from the same regions. As far as the blood vessels used are concerned, endogenous prostaglandin I_2 appears to modulate venous functions more than those of arteries, whereas endothelium-derived relaxing factor may be a more preferential modulator in the arteries than in the veins.

Key words: Artery—Acetylcholine—Histamine—Angiotensin II—Prostaglandin I_2—Nitric oxide—Mechanical response

Introduction

Mechanisms of action of various vasoactive substances have been analyzed pharmacologically in a variety of blood vessels. The necessity of studying diverse blood vessels is because the responses are heterogeneous in portions of vasculature, blood vessels of various organs and tissues, the vessels from different mammals, arteries and veins, and so on. The investigations so far reported are mainly on arteries and arterioles in relation to the control of blood pressure or blood flow and the genesis of vasospasm. Because of evidence showing contrasting responses of arteries and accompanying veins, recently, attention has focused on research into the venous system.

In this chapter, we concentrate on the effects of acetylcholine, histamine, angiotensin II, prostaglandin I_2 (PGI_2), and nitrovasodilators on isolated vein preparations, and compare the effects with those on arteries.

---

[1] Department of Pharmacology, Shiga University of Medical Sciences, Seta, Ohtsu, 520-21 Japan

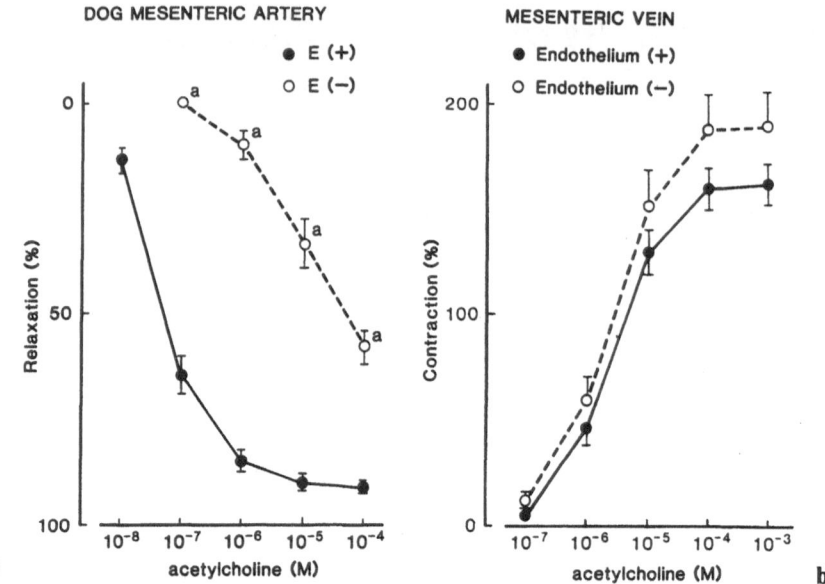

FIG. 1a,b. Concentration response curves of acetylcholine in dog mesenteric arterial (a) and venous (b) strips with $(E(+))$ and without $(E(-))$ endothelium. Contractions induced by $5\,mM$ Ba^{2+} were taken as 100% contraction for the veins, and relaxations induced by $10^{-4}\,M$ papaverine were taken as 100% relaxation for the arteries. A significant difference was found between arteries with and without endothelium; a, $P <$ 0.001. (From [3] and [7] with permission)

Acetylcholine

The addition of acetylcholine (10^{-7} to $10^{-4}\,M$) produces a dose-related contraction in dog mesenteric, pulmonary, femoral, and saphenous veins under resting conditions [1–3] and a relaxation in dog femoral, saphenous, and splenic veins [2] and monkey mesenteric veins (Toda et al., unpublished data) precontracted with vasoconstrictors. These responses are suppressed by atropine in a concentration-dependent manner. Contractions of dog pulmonary veins are potentiated by endothelium denudation [2], and relaxations of dog femoral veins are abolished by removal of the endothelium and treatment with hemoglobin or N^G-nitro-L-arginine, a nitric oxide (NO) synthase inhibitor [4,5], suggesting the involvement of endothelium-derived relaxing factor (EDRF) or NO in venous relaxation. In contrast, contractions of dog mesenteric veins are not dependent on the endothelium [3] (Fig. 1b), and relaxations of dog splenic [2] and monkey mesenteric veins (Toda et al., unpublished data) are not inhibited by endothelial damage. Therefore, acetylcholine, known as an EDRF-releasing substance [6], does not necessarily act on the endothelium to liberate relaxing substances in the vein.

On the other hand, dog mesenteric (Fig. 1a), pulmonary, femoral, saphenous, and splenic arteries, and monkey mesenteric arteries respond to acetylcholine with relaxations, which are always dependent on the endothelium. However, the response is not mediated exclusively by EDRF. For instance, the acetylcholine-induced monkey mesenteric arterial relaxation is associated partially with EDRF and also with PGI_2 released from subendothelial tissues [7].

Histamine

Histamine (10^{-7} to $10^{-4} M$) contracts rabbit veins and guinea-pig jugular and dog portal veins [8–11], and relaxes dog mesenteric veins [12], monkey mesenteric (Toda et al., unpublished data) and pulmonary veins [13], and rat jugular veins [9], all precontracted with vasoconstrictors. The venous contraction is reversed to a relaxation by treatment with histamine H_1 receptor antagonists, such as mepyramine and chlorpheniramine. Inositol triphosphate acts as an intracellular messenger for arterial contraction [14]; however, whether this is also the case for veins has not been determined.

Relaxations induced by histamine are inhibited by metiamide or cimetidine in the rat jugular vein, rabbit veins, dog mesenteric vein, and monkey pulmonary vein [9,12,13,15], suggesting the involvement of H_2 receptors. According to Matsuki and Ohhashi [13], the relaxant response is attenuated by H_1 and H_2 receptor antagonists and abolished by combined treatment with these antagonists, as is demonstrated in human and monkey cerebral and coronary arteries [16–18]. Therefore, both receptor subtypes are supposed to participate in the relaxation. In monkey pulmonary veins, activation of the H_1

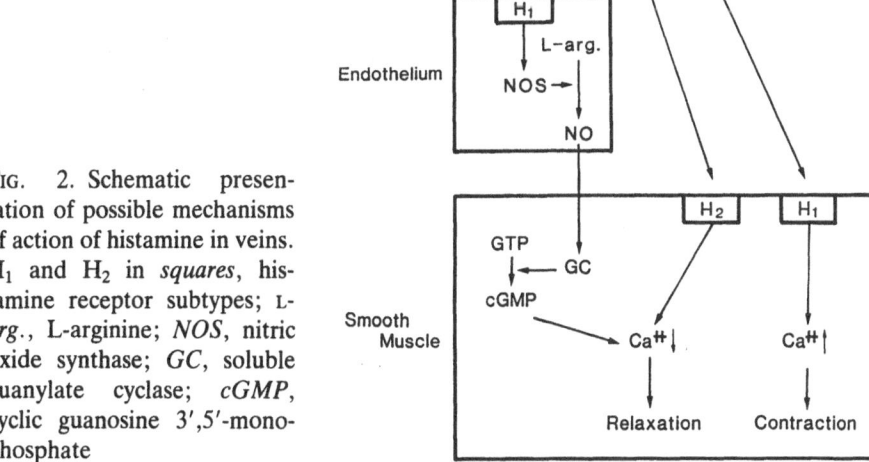

FIG. 2. Schematic presentation of possible mechanisms of action of histamine in veins. H_1 and H_2 in *squares*, histamine receptor subtypes; L-arg., L-arginine; *NOS*, nitric oxide synthase; *GC*, soluble guanylate cyclase; *cGMP*, cyclic guanosine 3′,5′-monophosphate

receptor subtype in the endothelium appears to be involved in the synthesis and release of EDRF, that in turn activates soluble guanylate cyclase, resulting in the production of cyclic GMP. Possible mechanisms of histamine action in veins are summarized in Fig. 2.

In dog mesenteric arteries, histamine-induced relaxation is associated with activation of the H_2 subtype and with the release of PGI_2 by stimulation of H_1 receptors in the endothelium [12].

Angiotensin II

Dog mesenteric venous strips precontracted with prostaglandin $F_{2\alpha}$ ($PGF_{2\alpha}$) respond to angiotensin II with a slight or no contraction followed by a moderate relaxation, whereas dog mesenteric arterial strips contract in response to the octapeptide [19]. The dose-response curves are compared in Fig. 3. The mean values of the apparent median effective concentration (EC50) in the veins and arteries are almost identical (1.5 and $1.2 \times 10^{-8} M$, respectively). The responses are abolished by treatment with angiotensin II receptor antagonists, such as saralasin and losartan.

The peptide-induced venous relaxation is independent of the endothelium and reversed to a contraction by treatment with cyclooxygenase inhibitors (Fig.

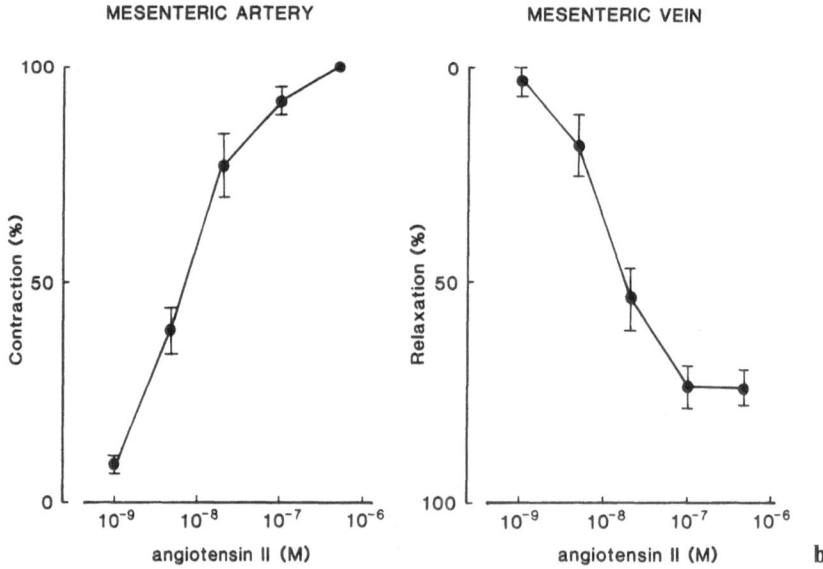

FIG. 3a,b. Concentration response curves of angiotensin II in dog mesenteric arteries (a) and veins (b). Contractions induced by $30\,mM$ K^+ were taken as 100% contraction for the arteries, and relaxations induced by $10^{-4} M$ papaverine were taken as 100% relaxation for the veins. (From [19] with permission)

FIG. 4. Modification by $10^{-6}M$ indomethacin of the response to angiotensin II ($2 \times 10^{-8}M$) of dog mesenteric arterial (*lower panel*) and venous (*upper panel*) strips. The venous strip was partially contracted with prostaglandin $F_{2\alpha}$ ($PGF_{2\alpha}$); *horizontal lines* just *left* of each tracing represent the level prior to the addition of $PGF_{2\alpha}$. *PA* represents $10^{-4}M$ papaverine that produced the maximal relaxation. (From [19] with permission)

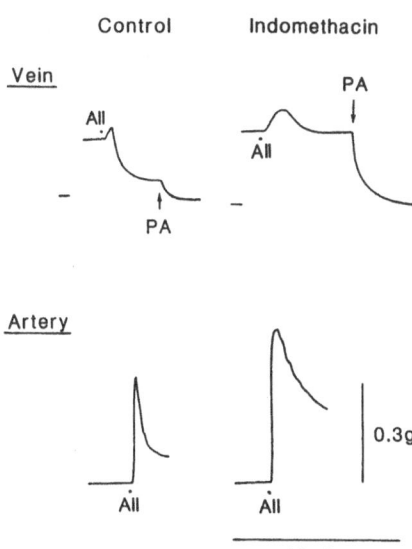

4). The arterial contraction caused by the peptide is potentiated by indomethacin and aspirin. The amount of 6-keto $PGF_{1\alpha}$, a stable metabolite of PGI_2, in the bathing media containing veins or arteries is increased by the addition of angiotensin II [20], suggesting the stimulated release of PGI_2 from the tissues. Contrasting responses to the octapeptide of dog mesenteric venous and arterial strips are not due to different mechanisms but to a different balance between the contractile and relaxant responses. In the veins, relaxations mediated by PGI_2 are predominant over the contractions, and vice versa in the arteries.

Arachidonic Acid and Prostaglandin I_2

Dog mesenteric venous strips partially contracted with $PGF_{2\alpha}$ respond to arachidonic acid (AA, 10^{-8} to $10^{-6}M$) with a dose-dependent relaxation [19]. The apparent EC50 value is $6.4 \times 10^{-8}M$, and the maximal relaxation relative to that caused by $10^{-4}M$ papaverine is 96%. The relaxation is markedly attenuated or abolished by treatment with indomethacin. Among available prostaglandins and thromboxane A_2 analogs, consistent, evident relaxation is induced only with PGI_2. Therefore, the AA-induced relaxation, susceptible to cyclooxygenase inhibitors, appears to be associated with PGI_2 synthesized in the venous wall. The addition of PGI_2 (10^{-9} to $10^{-6}M$) produces mesenteric venous relaxation; the EC50 averages $5.0 \times 10^{-8}M$, and the maximal relaxation is 98%.

In dog mesenteric arterial strips, the apparent EC50 value of AA ($2.0 \times 10^{-7}M$) is significantly greater than that in the venous strips, whereas the relaxations induced by PGI_2 are almost identical in these preparations [19]. More production of PGI_2 from AA in venous tissues than in arterial tissues is postulated.

Ca^{2+} Ionophore A23187, Nitroglycerin, and Nitric Oxide

The addition of Ca^{2+} ionophore A23187 (10^{-8} and $10^{-7} M$) relaxes dog mesenteric venous strips dose-dependently. The relaxation is not influenced by indomethacin but is reversed to a contraction by endothelium denudation [19]. Relaxations of dog femoral veins are also dependent on the endothelium and attenuated by N^G-nitro-L-arginine [5]. EDRF (NO) would participate in this response, which is not influenced by methylene blue, an inhibitor of soluble guanylate cyclase [21]. Therefore, the relaxation is supposed to be mediated by a mechanism distinct from changes in cyclic GMP [22]. Nitroglycerin (10^{-9} to $10^{-6} M$) and nitric oxide (10^{-8} to $10^{-4} M$), applied as acidified $NaNO_2$ solution [23], elicit the venous relaxation.

Relaxations induced by nitroglycerin and nitric oxide in dog mesenteric venous and arterial strips do not significantly differ. Greater responsiveness to nitroglycerin in veins than in arterioles has been postulated in vivo [24,25]. As far as the responses of isolated dog mesenteric veins and arteries are concerned, this is not the case, probably because the comparison was not made in the vein and arteriole/resistance vessel.

Conclusion

Responses of vein preparations to acetylcholine, histamine, angiotensin II, and other vasodilators, and the mechanisms of their actions, were compared with those for arteries. Even from these limited data, we can recognize that responses and the mechanisms of action differ among veins from different organs and tissues and from different mammals. From the results so far, it appears that veins are less able than arteries to liberate EDRF in response to activation of drug receptors, whereas the ability of veins to release PGI_2 appears to be similar to or greater than that of arteries. How characteristic features in the responsiveness to chemical stimuli of venous smooth muscle contribute to the regulation of venous capacitance and circulating blood volume might be determined by systematic, correlated analyses of venomotor responses in vitro and in vivo.

References

1. Ishikawa N, Ichikawa T, Shigei T (1980) Possible embryogenetical differences of the dog venous system in sensitivity to vasoactive substances. Jpn J Pharmacol 30:807–818

2. De Mey JG, Vanhoutte PM (1982) Heterogenous behavior of the canine arterial and venous wall: Importance of the endothelium. Circ Res 51:439–447
3. Miyazaki M, Toda N (1986) Endothelium-dependent changes in the responses to vasoconstrictor substances of isolated dog mesenteric veins. Jpn J Pharmacol 42:309–316
4. Miller VM, Vanhoutte PM (1989) Is nitric oxide the only endothelium-derived relaxing factor in canine femoral veins? Am J Physiol 257:H1910–H1946
5. Miller VM (1991) Selective production of endothelium-derived nitric oxide in canine femoral veins. Am J Physiol 261:H677–H682
6. Furchgott RF (1983) Role of endothelium in responses of vascular smooth muscle. Circ Res 53:557–573
7. Okamura T, Minami Y, Toda N (1989) Endothelium-dependent and -independent mechanisms of action of acetylcholine in monkey and dog isolated arteries. Pharmacology 38:279–288
8. Cook DA, Macleod KM (1978) Responses of rabbit portal vein to histamine. Br J Pharmacol 62:165–170
9. Cohen ML, Wiley KS (1978) Rat jugular vein relaxes to norepinephrine, phenylephrine and histamine. J Pharmacol Exp Ther 205:400–409
10. Tsuru H, Iwata M, Shigei T (1983) Relaxation of isolated rabbit veins mediated by latent histamine H_2-receptors. Experientia 39:577–578
11. Toshimitsu Y, Uchida K, Kojima S, Shimo Y (1984) Histamine responses mediated via H_1- and H_2-receptors in the isolated portal vein of the dog. J Pharm Pharmacol 36:404–405
12. Yamazaki M, Toda N (1992) Mechanisms of histamine-induced relaxation in isolated dog mesenteric arteries and veins. Folia Pharmacol Japon 99:19–26 (Abstract in English)
13. Matsuki T, Ohhashi T (1990) Endothelium and mechanical responses of isolated monkey pulmonary veins to histamine. Am J Physiol 259:H1032–H1037
14. Marche P, Girard A (1988) Phosphoinositides and cicletanine. Drugs Exp Clin Res 14:103–108
15. Tsuru H, Kohno S, Iwata M, Shigei T (1987) Characterization of histamine receptors in isolated rabbit veins. J Pharmacol Exp Ther 243:696–702
16. Toda N (1986) Mechanisms of histamine-induced relaxation in isolated monkey and dog coronary arteries. J Pharmacol Exp Ther 239:529–535
17. Toda N (1987) Mechanism of histamine actions in human coronary arteries. Circ Res 61:280–286
18. Toda N (1990) Mechanism underlying responses to histamine of isolated monkey and human cerebral arteries. Am J Physiol 258:H311–H317
19. Yamazaki M, Toda N (1991) Comparison of responses to angiotensin II of dog mesenteric arteries and veins. Eur J Pharmacol 201:223–229
20. Yoshida K, Yamazaki M, Toda N (1991) Different modulation by cyclooxygenase inhibitors of the response to angiotensin II in monkey arteries and veins. Jpn J Pharmacol 55:469–475
21. Gruetter CA, Kadowitz PJ, Ignarro LJ (1981) Methylene blue inhibits coronary arterial relaxation and guanylate cyclase activation by nitroglycerin, sodium nitrite and amyl nitrite. Can J Physiol Pharmacol 59:150–156
22. Vidal M, Vanhoutte PM, Miller VM (1991) Dissociation between endothelium-dependent relaxations and increases in cGMP in systemic veins. Am J Physiol 260:H1531–H1537

23. Furchgott RF (1988) Studies on relaxation of rabbit aorta by sodium nitrite: the basis for the proposal that the acid-activatable inhibitory factor from bovine retractor penis is inorganic nitrite and the endothelium-derived relaxing factor is nitric oxide. In: Vanhoutte PM (ed) Vasodilatation. Raven, New York, pp 401–414
24. Mason DT, Braunwald E (1965) The effects of nitroglycerin and amyl nitrite on arteriolar and venous tone in the human forearm. Circulation 32:755–766
25. Gilman AG, Rall TW, Nies AS, Taylor P (1990) The pharmacological basis of therapeutics. Pergamon, New York

Coronary Venous Flow

FUMIHIKO KAJIYA, AKIHIRO KIMURA, OSAMU HIRAMATSU, YASUO OGASAWARA, and KATSUHIKO TSUJIOKA[1]

Abstract. It is well known that coronary arterial flow is predominantly diastolic, whereas coronary venous flow is predominantly systolic. Since coronary venous flow is squeezed out from the myocardial vascular bed by direct and indirect extravascular compressive forces of the myocardium, the coronary venous system is therefore very unique and offers a suitable model to investigate the relationship between cardiac contraction and coronary blood flow. Moreover, the intramyocardial coronary venous system has a negative feedback control system against arterial inflow into the myocardium; that is the increase of intramyocardial venous blood volume decreases arterial inflow and this decrement enhances arterial inflow. Therefore, analysis of coronary venous outflow in relation to the mechanical control of coronary arterial inflow is also important. We overviewed the phasic blood flow patterns of proximal and distal coronary vessels of the left ventricle, right ventricle, and left atrium.

In summary, the interaction at the microscopic level between myocytes and microvessels may be much the same in both ventricles and the left atrium.

Key words: Coronary venous flows—Laser Doppler velocimeter—Myocardial contractility

Introduction

More than 300 years ago, in 1689, Scaramucci hypothesized that the deeper coronary vessels are squeezed by contraction of the muscle fibers around them, which displaces the intramyocardial blood into coronary veins, and that the vessels are refilled from the aorta during diastole [1].

---

[1] Department of Medical Engineering and Systems Cardiology, Kawasaki Medical School, 577 Matsushima, Kurashiki, 701-01 Japan

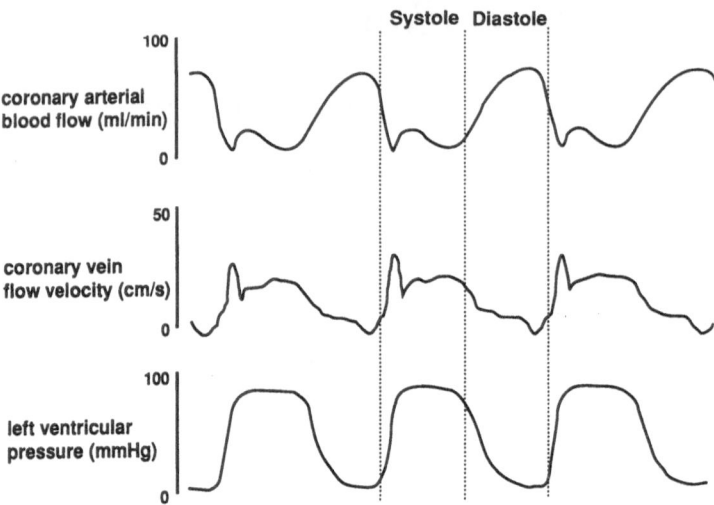

FIG. 1. Simultaneous recordings of coronary arterial blood flow, coronary small vein flow velocity and left ventricular pressure. Coronary artery flow into the left ventricular myocardium is almost exclusively diastolic, whereas coronary venous blood is squeezed out from the myocardium mainly during systole

To prove this hypothesis, many investigators have studied the relationship between cardiac contraction and blood flows in coronary arteries and veins. However, no significant progress was made on the relationship between cardiac contraction and the coronary blood flows during the 18th and 19th centuries [2–4]. Following the development of better measuring techniques for coronary arterial and venous flows during this century, the characteristics of coronary flow were investigated in relation to cardiac contraction and relaxation. Now, it is well documented that coronary arterial flow is almost restricted to diastole, whereas coronary venous flow is systolic (Fig. 1). Since coronary venous blood is squeezed out from the myocardial vascular bed by the direct extravascular compressive force of heart muscle, the coronary venous system is a suitable model to investigate the relationship between cardiac contraction and coronary blood flow.

Our laser Doppler velocimeter (LDV) with an optical fiber is a powerful device for the measurement of coronary venous flow [5]. The most important advantage of the device over conventional velocimeters is its excellent accessibility to the vessel, even when the vessels move with a cardiac cycle and they are easily deformable, as are coronary veins. In this chapter, following a brief description of our laser Doppler system, we report our results on these topics: (1) blood flow velocities in the coronary veins of the left ventricle; (2) the blood flow velocity waveform in small epicardial coronary veins of the right ventricle; and (3) the phasic blood flow velocity waveform of atrial small veins.

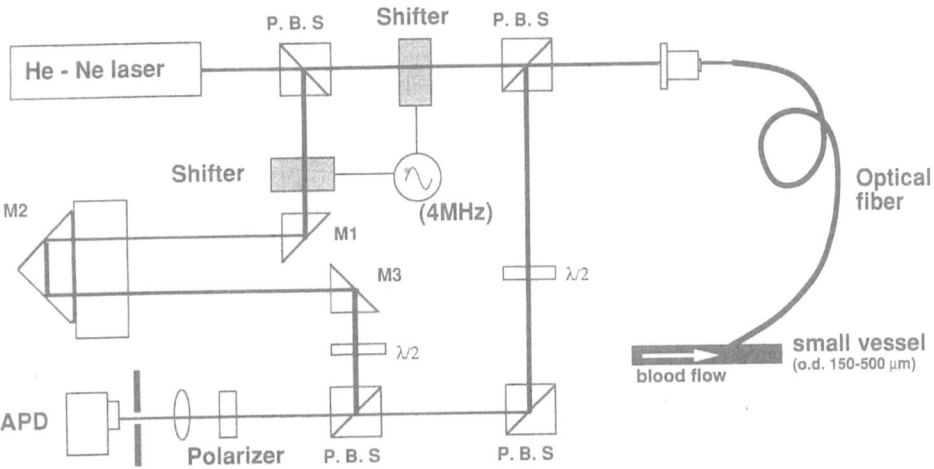

FIG. 2. Schematic diagram of the laser Doppler velocimeter with an optical fiber (see text for details). Mirrors 1, 2, and 3 (*M1*, *M2*, and *M3*) were used to get an optimal path-length of laser beam by changing the position of mirror 2. *P.B.S.*, polarizing beam splitter; *A.P.D.*, avalanche photo diode; *o.d.*, outer diameter. (From [5] with permission from Birkhauser Verlag)

A Laser Doppler Velocimeter with an Optical Fiber

Measurements of blood flow velocity in coronary veins were performed with a laser Doppler velocimeter (LDV) incorporating an optical fiber, as previously described in detail by Kajiya et al. [5] and Nishihara et al. [6]. Briefly, a He-Ne laser beam (wave length, 632.8 nm and power, 5 mW) is divided into incident and reference beams by a beam splitter (P.B.S.; Fig. 2). The incident beam is directed onto the vascular surface through an optical fiber (external diameter: 62.5 μm and core diameter: 50 μm) and then introduced into the vascular lumen through the vessel wall. Part of the light back-scattered through the vascular wall by flowing erythrocytes is collected by the same fiber and transmitted back. The other light divided by the beam splitter is used as the reference beam. A frequency shifter (4 MHz) is interposed in the path of the reference beam to differentiate forward from retrograde flow. The photocurrent from an avalanche photodiode (APD) is fed into a spectrum analyzer to detect Doppler shift frequencies. The maximum Doppler shift frequency in the sample volume—that is, the maximum velocity—is detected automatically:

$$V = \Delta f \cdot l / 2n\cos\theta \qquad (1)$$

where Δf is the Doppler shift frequency, l is the laser wavelength (632.8 nm) and n is the refractive index of blood (approximately 1.33). The angle θ between the fiber and the vascular axis was measured after experiments

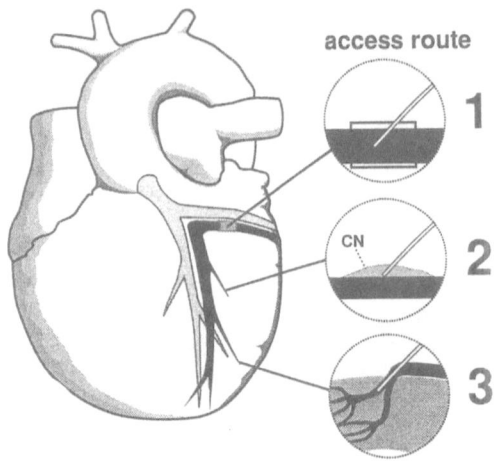

access route

FIG. 3. Three different routes of access of the fiber probe to coronary vessels. Route 1 was applied for velocity measurements in large and middle-sized epicardial coronary veins, e.g., the great cardiac vein. Route 2 was for velocity measurements in small epicardial veins; *CN*, cyanoacrylate. Route 3 was for velocity measurements in intramyocardial veins (see text for details)

employing a protractor to calculate blood velocities using Eq. 1. To test the accuracy of our method, we compared the blood velocities measured by our LDV with the blood flow rate determined by timed collection, assuming that velocity profiles across the vessel are parabolic [7]. The correlation coefficient was 0.96–0.99, indicating that our method accurately measures blood flow velocities.

For the blood flow velocity measurements in coronary veins, we used three different routes of access for the fiber probe according to the measuring sites (Fig. 3), i.e., epicardial coronary veins, epicardial small veins (outer diameter <1.0 mm), and intramyocardial vessels. First (access route 1), for velocity measurement in a relatively large coronary vein, we placed a cuff around the vein and inserted the fiber into the vessel through a small hole in the vessel wall. Second (access route 2), for velocity measurement in an epicardial small vein in which the wall was thin enough to be transparent for laser light, we placed the fiber tip on the vessel surface and fixed it by a drop of cyanoacrylate glue. Third (access route 3), for velocity measurement in intramyocardial vessels, we inserted the fiber probe into the vascular lumen from a position penetrating into myocardium and introduced the fiber into an intramyocardial coronary vessel.

Blood Flow Velocities in Epicardial Large and Small Coronary Veins and Intramyocardial Veins of the Left Ventricle

For the blood flow velocity measurements in large epicardial veins, we placed a cuff around the vessel and inserted the fiber into the vessel through a small hole in the vessel wall (access route 1, see Fig. 3). A representative tracing of

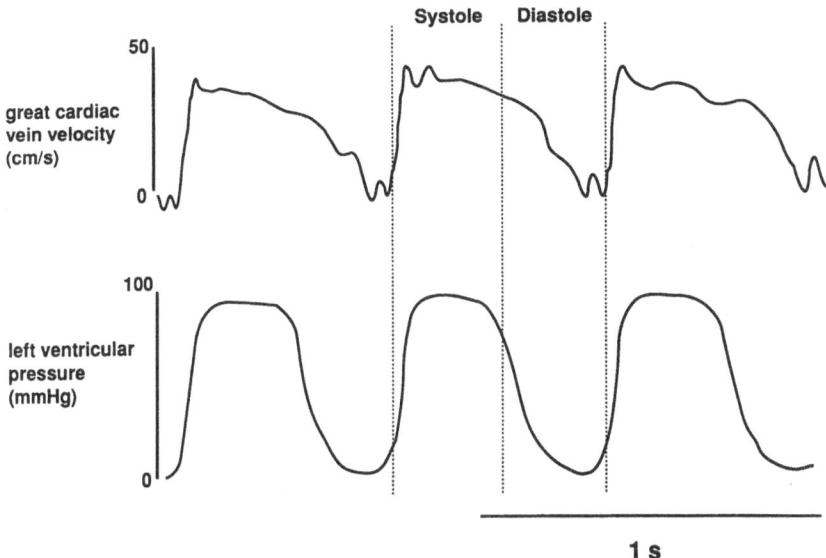

Fᴵɢ. 4. Phasic blood velocity waveform near or on the central axial region of the great cardiac vein. The velocity waveform was characterized by a prominent systolic flow wave. (Modified from [8] with permission from Springer-Verlag)

blood flow velocity in the large epicardial vein, i.e., great cardiac vein (GCV), is shown in Fig. 4 [8]. The GCV velocity is always characterized by a predominantly systolic flow wave. The blood velocity increased around the onset of left ventricular ejection and decreased gradually after the peak formation at mid- or late-systole. These findings are consistent with earlier observations made using flow-velocity or volumetric-flow measuring systems [9–12]. Recently, Canty and Brooks [10] also found volumetric coronary sinus flow to be predominantly systolic. However, these authors suggested that there may be a significant effect of coronary venous compliance on the measurements of phasic coronary venous flow which could obscure the venous outflow from myocardium, because the large epicardial veins act as capacitors. Therefore, we measured the blood velocity in a small epicardial vein just after its appearance from myocardium in the dog. Vessels with an outer diameter of about 150–500 μm were chosen for the observation. We used access routes 2 and 3 shown in Fig. 3 for the velocity measurements of small coronary veins. When large and good-quality Doppler signals were consistently observed, the fiber tip was fixed at that position with a drop of cyanoacrylate glue.

The left panel of Fig. 5 shows the blood velocity pattern in a small epicardial coronary vein by access route 2, and also the velocity pattern in the intra-myocardial small vein by access route 3 (approximately 2 mm beneath the cardiac surface). In both cases, blood flow was predominantly systolic, as seen in the great cardiac vein and coronary sinus. However, the onsets of acceleration

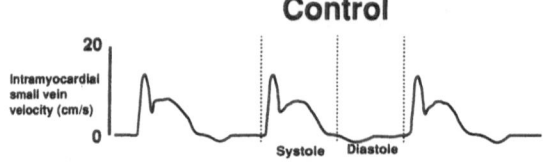

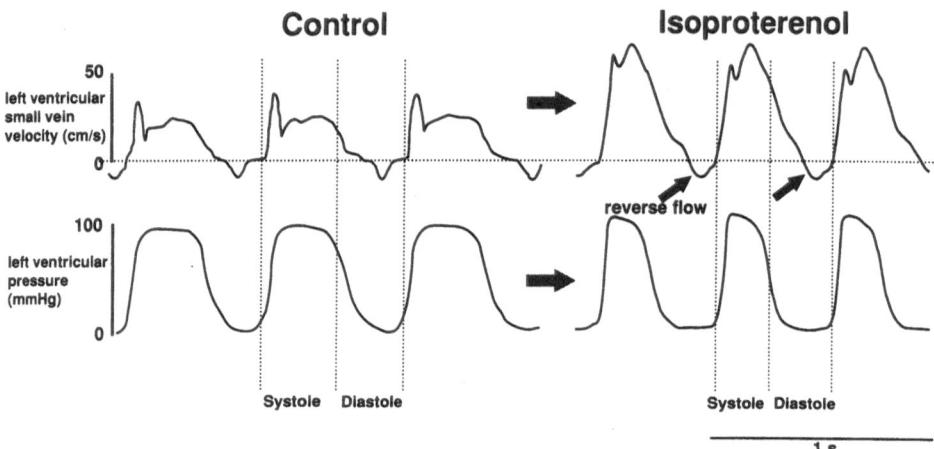

FIG. 5. Phasic blood velocity waveforms in the intramyocardial small vein as measured by access route 3 (*left upper panel*), and a left ventricular small vein by access route 2 (*left lower panel*). Both exhibited similar waveforms with a predominantly systolic flow. Note the existence of diastolic retrograde flows. Increase in contractility by isoproterenol (ISP) administration (*right panel*) increased the peak venous blood flow velocity, and also accelerated the rise in the initial systolic flow wave. The diastolic reverse flow was also augmented by ISP

and deceleration of the flow in both the small epicardial and the intramyocardial veins were earlier than those in the large epicardial vein; i.e., the velocity increased with the ventricular isovolumic contraction phase and decreased rapidly with beginning of ventricular relaxation. A reverse flow was observed during the early part of diastole in most cases. The increase in contractility generated by isoproterenol (ISP) administration raised the peak venous blood flow velocity by about 140%, and accelerated the rise in the systolic flow velocity. A diastolic reverse flow was also augmented by increased cardiac contractility (see the right panel of Fig. 5). Left ventricular cavity pressure before and after ISP administration was not changed significantly. Thus, increased myocardial contractility augments the myocardial squeezing-out effect on the intramyocardial venous blood, and also the diastolic myocardial suction force on the blood pooled in epicardial capacitance vessels.

Figure 6 shows the phasic blood velocity pattern of an epicardial small vein before and during aortic constriction. Although the left ventricular pressure

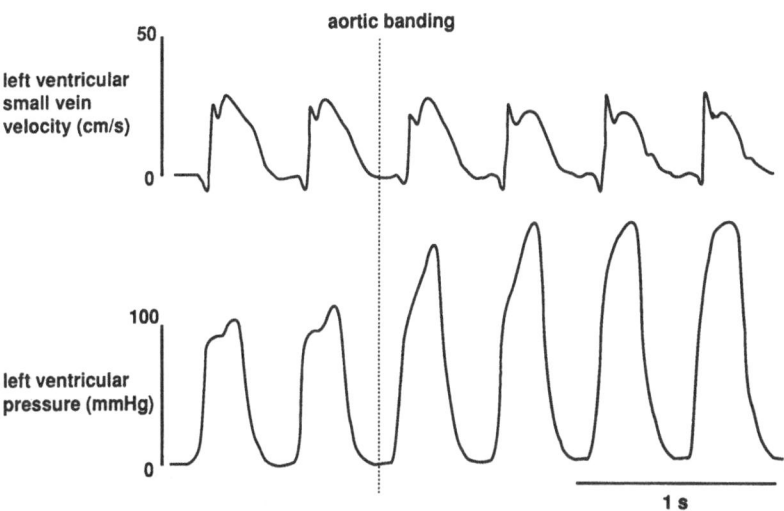

FIG. 6. Typical tracing of the phasic blood velocity pattern of an epicardial small vein before and during aortic constriction. Although left ventricular pressure was increased by about 150% during aortic constriction, the phasic blood velocity pattern of the epicardial small vein was not affected significantly

was increased 150% by aortic constriction, the phasic blood velocity pattern in the epicardial small vein was not affected significantly. These results indicate that the myocardial contractility is one of the major determinants of phasic coronary venous flow, and that ventricular cavity pressure only weakly correlates with the phasic pattern of coronary venous flow.

As for the phasic diameter change in venules, Nellis et al. [13] found that the diameter of subendocardial venules increased in end-systole and/or early-diastole by about 20% as measured with a free-motion technique, whereas we observed that the diameter of subendocardial venules decreased at end-systole by about 20% as measured with a needle-probe video-microscope with a charge-coupled device (CCD) camera [14]. There are two factors which may contribute to this difference. First, the systolic increase in venous outflow from deeper myocardial layers to the epicardial veins, together with some resistance within the epicardial veins, may result in an increase in intraluminal venous pressure and hence an increase in epicardial and subendocardial venular diameter during systole. Second, the extravascular compressive force associated with left ventricular pressure may be weaker in subepimyocardium than that in subendomyocardium. Therefore, although the effect of myocardial systole on subendocardial and subepicardial vessels may be similar, the difference in extravascular compressive force caused by left ventricular pressure may result in transmural variations of extravascular compressive force.

In summary, blood in the coronary veins of the left ventricle is squeezed out from the myocardial vascular bed by the extravascular compressive force mainly due to heart muscle contraction, but transmural variation of vein flows may be affected by the intramyocardial pressure due to left ventricular pressure.

Blood Flow Velocity in Small Epicardial Coronary Veins of the Right Ventricle

The venous outflow from small right ventricular veins has not been previously studied, although blood velocity measurements in the right ventricular veins are crucial for the assessment of the relationship between intramyocardial venous outflow and cardiac contraction. To clarify the characteristics of their phasic blood-flow waveforms, we measured blood velocities in peripheral portions of the right coronary veins (outer diameter of about 150–500 μm) in dogs under the following three conditions: (1) control, (2) isoproterenol (ISP) administration, and (3) nitroglycerin administration [15]. Measurements of blood flow velocity were performed with our laser Doppler velocimeter with an optical fiber, using access route 2 (see Fig. 3).

Figure 7 shows a tracing of the blood velocity in a small vein of the right ventricle just after its appearance from myocardium. The velocity waveform in

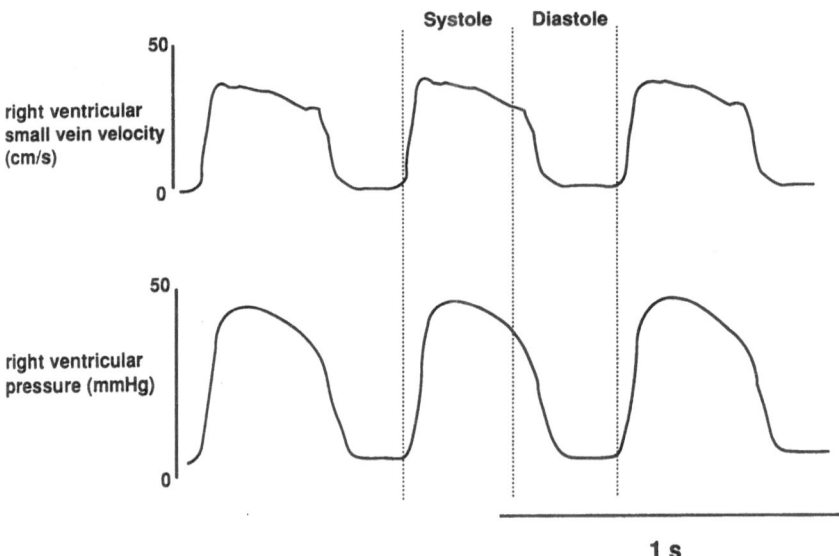

FIG. 7. The blood velocity waveform in an epicardial small vein of the right ventricle. The blood velocity waveform was characterized by predominantly systolic flow, a pattern which is basically the same as for the left ventricular venous flow. (From [15] with permission from the American Physiological Society)

small right coronary veins was predominantly systolic, i.e., it increased with a rise in right ventricular pressure and decreased with right ventricular relaxation. The peak velocity under control conditions was 16.0 ± 2.1 cm/s. ISP administration increased the acceleration rate of the venous velocity and increased the peak velocity by 38% ($P < 0.01$) compared to control conditions. This increased acceleration of the venous outflow indicated that myocardial contractility contributes to the phasic nature of the venous flow from the right ventricular myocardium. Nitroglycerin administration increased the peak velocity by 40% ($P < 0.01$) compared to control conditions, but it did not increase flow acceleration in early systole. Aortic pressures were decreased significantly following ISP or nitroglycerin administration, but the changes in heart rate were not significant. Thus, the increase in the peak systolic venous flow velocity following nitroglycerin may be mainly caused by an increase during diastole of blood volume stored in the intramyocardial capacitance vessels. In contrast, the flow increase following ISP may be mainly caused by augmented cardiac muscle contractility.

In summary, increases in both the myocardial contraction of the right ventricle and the intramyocardial blood volume promote venous outflow from the intramyocardial capacitance vessels. The interaction on the microscopic level between myocytes and microvessels may be much the same in the right and left ventricle.

Phasic Blood Flow Velocity of Atrial Small Vessels

It is worth studying the phasic velocity waveform of atrial bood flow, since atrial muscle contraction generates much less atrial cavity pressure than left ventricular pressure. However, there have been few studies on the phasic patterns of atrial flow because of the technical difficulties of measurement. Recently, Kajiya et al. [7] measured the atrial small arterial and venous blood flow velocity. They reported that in small arteries, blood velocity showed a waveform with a pattern similar to that of aortic (coronary perfusion) pressure during atrial relaxation, and inflow was impeded by atrial contraction, whereas blood flow in small veins was observed mainly during atrial contraction (Fig. 8) [7]. This 180° phase difference between arterial and venous flows indicates that a large portion of the arterial flow into the atrial myocardium may be stored in the intramyocardial vessels. We estimated the left ventricular intramyocardial capacitance to be almost 0.1 ml/mmHg per 100 g LV myocardium by simultaneous measurement of coronary artery and vein flows in the dog [16]. In a similar way, we roughly estimated the blood volume stored in the atrial intramyocardial vessels to be about one-fourth of that in the left ventricular myocardium during ventricular diastole [7]. Then, the problem is the nature of the atrial myocardial compressive forces acting on the atrial capacitance vessels.

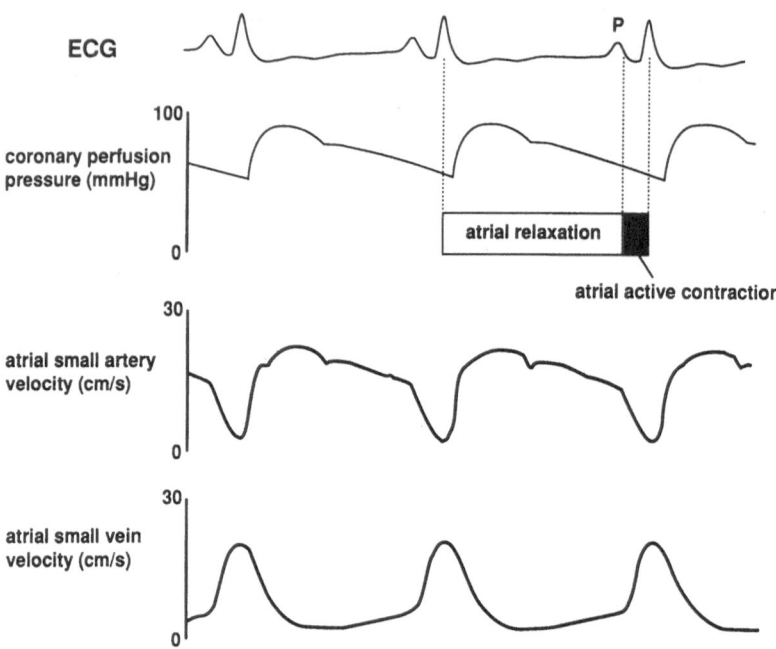

Fig. 8. Schematic presentation of the phasic blood velocity pattern in a left atrial small artery and vein. The arterial blood flow corresponded with the pattern of coronary perfusion pressure, but flow was impeded by atrial contraction. In contrast, the venous blood velocity increases during atrial contraction. There is a 180° phase shift in these blood flows. (Modified from [7] with permission from the American Heart Association)

We tested the relative importance of atrial chamber pressure to atrial muscle contractility in the genesis of pulsatile atrial arterial and venous blood velocity. The genesis of the former is closely related to passive atrial intramyocardial pressure, and of the latter, to the muscle contractility (elastance) [17]. We measured blood velocities in the atrial small arteries and veins (outer diameter: 150–500 µm) in anesthetized open-chest dogs ($n = 21$). The velocity sensor (optical fiber) was fixed on the vessel surface with a drop of cyanoacrylate glue when good-quality Doppler signals were consistently observed. Left atrial pressure (LAP) and the contractility of the left atrium were changed by premature ventricular contraction (PVC) and intracoronary administration of isoproterenol (ISP; 0.5 µg), respectively.

Figure 9 shows the effect of atrial chamber pressure on the velocity waveform of the left atrial small arteries and veins. PVC increased LAP significantly during arterial velocity measurements, from 8.1 ± 2.7 (SD) to 16.4 ± 1.3 mmHg, and during venous measurements from 8.2 ± 1.2 to 14.3 ± 3.7 mmHg. However, PVC did not change the blood velocity patterns, e.g., the maximum deceleration rate of the velocity wave in arteries during atrial systole and the maximum acceleration rate of the systolic velocity wave in veins.

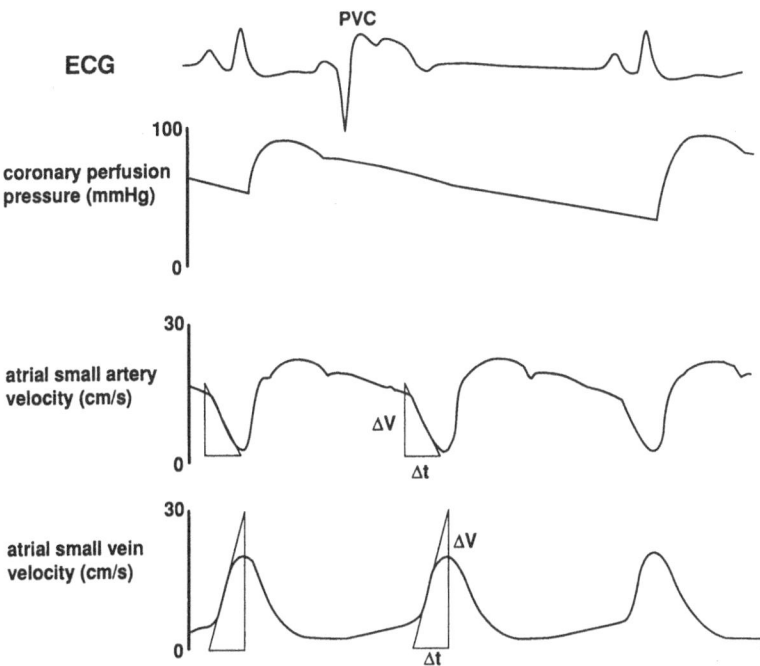

FIG. 9. A recording of left atrial small artery and vein velocities with ECG and coronary perfusion pressure. As in Fig. 8, the trace of the arterial blood velocity resembled the pattern of coronary perfusion pressure, but it was characterized by a pronounced dip (a sharp transient decrease in flow velocity) during atrial contraction. Premature ventricular contraction (PVC) doubled the atrial systolic pressure, but did not augment the impediment of arterial flow significantly. PVC did not affect the venous velocity waveform. (Modified from [17] with permission from the British Cardiac Society. $\Delta V \cdot \Delta t^{-1}$, rate of change of velocity)

Figure 10 shows the effect of atrial muscle contractility on the velocity waveform of the left atrial small arteries and veins. Although ISP did not change left atrial pressure, it decreased the minimum arterial blood velocity during atrial systole (from 3.3 ± 3.4 to $-2.5 \pm 3.2 \, \text{cm} \cdot \text{s}^{-1}$), and increased the maximum venous blood velocity (from 15.9 ± 5.5 to $19.2 \pm 7.4 \, \text{cm} \cdot \text{s}^{-1}$). ISP also increased both the maximum deceleration rate of the systolic velocity wave in arteries (from 90 ± 45 to $234 \pm 143 \, \text{cm} \cdot \text{s}^{-2}$) and the maximum acceleration rate of the velocity wave in veins during atrial systole (from 356 ± 30 to $763 \pm 366 \, \text{cm} \cdot \text{s}^{-2}$). These results indicate: (1) that left atrial pressure is not a major determinant of the blood flow patterns of atrial veins, and (2) that atrial contractility significantly affects the blood flow patterns of atrial veins, although the extravascular compressive forces caused by atrial pressure and atrial contractility may interact with each other.

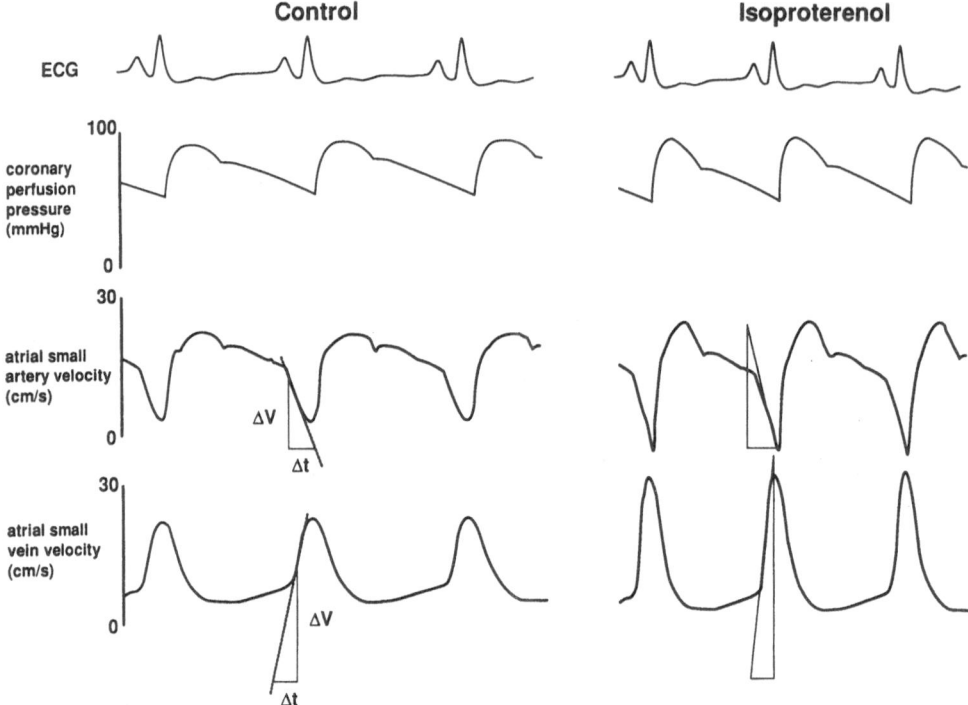

FIG. 10. Typical tracings of the blood velocities in a left atrial small artery and vein under control conditions (*left panel*) and ISP administration (*right panel*). $\Delta V \cdot \Delta t^{-1}$, rate of change of velocity. Note that retrograde flow is present during atrial contraction (*right panel*). ISP also increased the maximum venous velocity and the $\Delta V \cdot \Delta t^{-1}$. (Modified from [17] with permission from the British Cardiac Society)

Concluding Remarks

Although the peak cavity pressures in the left and right ventricles and left atrium are different, it was a common finding among both ventricles and atrium that phasic coronary venous blood is squeezed out during the myocardial contraction phase. This indicates that muscle contraction is an important force to propel the blood pooled in intramyocardial capacitance vessels into coronary veins. The observation that venular diameter increased during systole in subepicardium of the left ventricle, while it decreased during systole in subendocardium, may be explained by the transmural difference of intramyocardial pressure. Thus, muscle contraction may affect the overall nature of the phasic pattern of coronary vein flow, whereas intramyocardial pressure may explain the transmural difference of venous flow.

In an earlier study, we demonstrated that the coronary venous system has a negative feedback control system against arterial inflow, i.e., the increment of intramyocardial venous blood volume during a diastole decreases arterial inflow [18]. Therefore, analysis of coronary venous outflow in relation to cardiac contraction and relaxation is also very important to the understanding of the mechanical control of coronary arterial inflow.

References

1. Scaramucci J (1695) Theoremata familiaria viros eruditos consulentia de variis physico-medicis lucubrationibus juxta leges mecanicas. Apud Joannem Baptistam Bustum 70–81
2. Porter WT (1898) The influence of the heart beat on the flow of blood through the walls of the heart. Am J Physiol 1:145–163
3. Anrep GV, Cruickshank EWH, Downing AC, Subba RA (1927) The coronary circulation in relation to the cardiac cycle. Heart Bull 14:111–133
4. Wiggers CJ (1954) The interplay of coronary vascular resistance and myocardial compression in regulating coronary flow. Circ Res 2:271–279
5. Kajiya F, Hoki N, Tomonaga G, Nishihara H (1981) A laser-Doppler-velocimeter using an optical fiber and its application to local velocity measurement in the coronary artery. Experientia 37:1171–1173
6. Nishihara H, Koyama J, Hoki N, Kajiya F, Hironaga M, Kano M (1982) Optical fiber laser Doppler velocimeter for high-resolution measurement of pulsatile blood flow. Appl Optics 21:1785–1790
7. Kajiya F, Tsujioka K, Ogasawara Y, Hiramatsu O, Wada Y, Goto M, Yanaka M (1989) Analysis of the characteristics of the flow velocity waveforms in left atrial small arteries and veins in the dog. Circ Res 65:1172–1181
8. Kajiya F, Tsujioka K, Goto M, Wada Y, Tadaoka S, Nakai M, Hiramatsu O, Ogasawara Y, Mito K, Hoki N, Tomonaga G (1985) Evaluation of phasic blood flow velocity in the great cardiac vein by a laser Doppler method. Heart Vessels 1:16–23
9. Chilian WM, Marcus ML (1984) Coronary venous outflow persists after cessation of coronary artery inflow. Am J Physiol 247:H984–H990
10. Canty JM Jr, Brooks A (1990) Phasic volumetric coronary venous outflow patterns in conscious dogs. Am J Physiol 258:H1457–H1463
11. Klassen GA, Armour JA (1984) Coronary venous pressure and flow: effects of vagal stimulation, aortic constriction, ad vasodilators. Can J Physiol Pharmacol 62:531–538
12. Stein PD, Badeer HS, Schuette WH, Glaser JF (1969) Pulsatile aspects of coronary blood flow in closed-chest dogs. Am Heart J 78:331–337
13. Nellis SN, Whitesell L (1989) Phasic pressures and diameters in small epicardial veins of the unrestrained heart. Am J Physiol 257:H1056–H1061
14. Yada T, Hiramatsu O, Kimura A, Goto M, Ogasawara Y, Tsujioka K, Yamamori S, Ohno K, Hosaka H, Kajiya F (1993) In vivo observation of subendocardial microvessels of the beating porcine heart using a needle-probe video-microscope with a CCD camera. Circ Res 72:939–946

15. Hiramatsu O, Kimura A, Yada T, Yamamoto T, Ogasawara Y, Goto M, Tsujioka K, Kajiya F (1992) Phasic characteristics of arterial inflow and venous outflow of right ventricular myocardium in dogs. Am J Physiol 262:H1422–H1427
16. Kajiya F, Tsujioka K, Goto M, Wada Y, Chen XL, Nakai M, Tadaoka S, Hiramatsu O, Ogasawara Y, Mito K, Tomonaga G (1986) Functional characteristics of intra-myocardial capacitance vessels during diastole in the dog. Circ Res 58:476–485
17. Kimura A, Hiramatsu O, Wada Y, Yada T, Yamamoto T, Goto M, Ogasawara Y, Tsujioka K, Kajiya F (1992) Atrial contractility affects phasic blood flow velocity of atrial small vessels in the dog. Cardiovasc Res 26:1219–1225
18. Goto M, Tsujioka K, Ogasawara Y, Wada Y, Tadaoka S, Hiramatsu S, Yanaka M, Kajiya F (1990) Effect of blood filling in intramyocardial vessels on coronary arterial inflow. Am J Physiol 258:H1042–H1048

Macromolecular Permeability and Hydraulic Conductivity Through Large Pores Across a Single Venular Capillary

Akira Kamiya, Masahiro Shibata, and Mitra Sohirad[1]

Abstract. To determine the hydraulic conductivity and permeability of macro-molecules across capillary and venular walls in the rabbit tenuissimus muscle, we developed a new fluorescence vital microscope in which a pair of well-defined slit laser beams ($10–50\,\mu$m in thickness) were focused, so as to cross in a chosen depth of the tissue region and to excite the fluorescent tracer intensively at the crossing portion adjusted to the focal plane of the objective. After injecting a fluorescent tracer (FITC-Dextran: FD40, 70, 150; M.W. = 40000, 70000, 150000) via the jugular vein, temporal changes in the tracer concentration within the cross-illuminated portion of the muscle tissue were recorded as fluorescent intensity alterations. We calculated the diffusive (D) and convective (V) components of the extra-vasated tracer by best-fitting the experimental curve with the simulation model based on the "Pore theory", in which macromolecules are assumed to pass across the vessel wall through the "Large pore" passively buy diffusion and convection. Diffusion permeability (P_d) and hydraulic conductivity (L_p) were determined from D and V values obtained at the centre and venular end of capillaries to compare their differences. The values of P_d at the venular ends were approximately 10 times greater than those at the centre but V values at the venular ends were only about two-fold greater than those of the central capillary. The mechanism bringing about such discrepancy was discussed.

Key words: Capillary—Venule—Permeability—Hydraulic conductivity—Diffusion—Convection—Large pore

Introduction

The major role of the venous system has conventionally been considered to be stabilization of the entire cardiovascular system as the capacitance vessel. However, recent evidence suggests that the veins, particularly the venules or

[1] Institute of Medical Electronics, University of Tokyo School of Medicine, 7-3-1 Hongo, Bunkyo-ku, Tokyo, 113 Japan

venular ends of the capillaries, carry out various other activities such as being the special site initiating the capillary angiogenesis [1], the preferential site of leukocyte migration across the vessel wall into the interstitial space [2], the dominant site for transcapillary exchange of plasma proteins or other macromolecules [3], etc. In this article, the macromolecular permeability across the venular wall will be described in terms of the diffusional and convectional transport components, measured at a single capillary in the rabbit skeletal muscle, by using a newly developed fluorescent vital microscope system.

Method

The System for Measuring the Local Permeability Across a Single Capillary

The mechanism of transport of macromolecules such as plasma proteins across microvessels has not yet been established. Extensive analyses of transcapillary permeability over the past 30 years, using data based on whole organ measurements, have not clarified whether the physical mechanisms—diffusion and convection through large pores (gaps) at the intracellular junctions of capillary endothelial cells—or the biological mechanism—cytoplasmic vesicular transport—plays the major role in macromolecular translocation [4]. The application of the double-pore theory [5] and Patlack's equation (see Eq. 3) in the analysis of steady-state solute flux assuming a physical transport mechanism [6] have posed many difficult problems for theoretical treatment of the experimental data. The heterogeneity of capillary permeability has also introduced various ambiguities into the analysis [7]. Hence, it is now widely accepted that we have to provide a new technique to measure macromolecular permeability directly at a single capillary in the in situ microcirculation. Such a technique would also be useful for measuring any local differences in permeability along a capillary channel.

In order to measure permeability in a single capillary, it is necessary to quantify precisely the tracer concentration in the tissue surrounding each capillary. Figure 1 illustrates a newly developed fluorescence vital microscope provided with a cross-illumination system of two very thin square laser beams of around 25 μm in thickness. The focal plane of the objective is automatically adjusted to the zone of intersection of the beams. Hence, when a fluorescent tracer is permeated into the tissue after its intravenous injection, the fluorescent light excited by the laser beam in the intersection zone alone can be imaged through the microscope. By focusing the objective on a single capillary at a certain depth of the tissue, we can obtain the image of the tracer concentration in the capillary and in the tissue surrounding it, with no interference from the surroundings of the intersection zone. Since the focal plane of the objective lens can be moved down vertically about 200 μm from

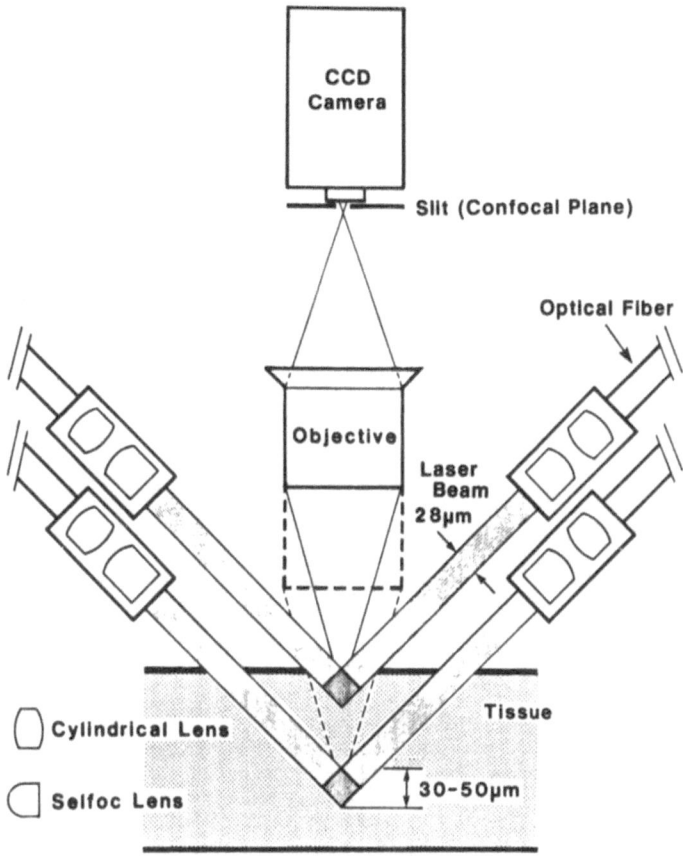

FIG. 1. A schematic illustration of a cross-illumination system with two very thin laser beams (25 μm in thickness), used in this study for microscopic measurement of fluorescent tracer permeability across a single capillary. Since the intersection point of the laser beams is optically adjusted to the focal plane of the objective lens, it is also possible to construct a three-dimensional image of microvascular architecture by moving the point horizontally (see text for details). *CCD*, computer-controlled display

the tissue surface, it is also possible to make a three-dimensional mapping of topological architecture of the in vivo microvasculature.

Figure 2 schematically illustrates the measuring system used to determine the permeability parameters of a single capillary in the rabbit tenuissimus muscle. Fluorescein isothiocyanate (FITC)-dextrans of molecular weights 40 000, 70 000 and 150 000 (FITC-Dx 40, 70 and 150) and of molecular radii 4.37, 5.79 and 8.25 nm respectively, were used as the fluorescent tracers. The time course of tissue concentration of these tracers around a single capillary after the

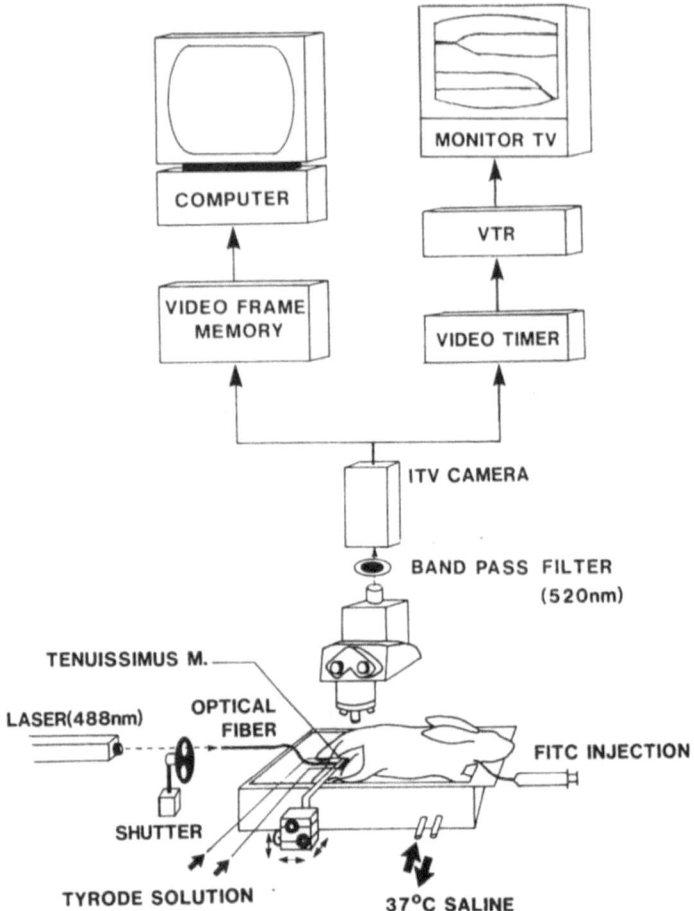

FIG. 2. A schematic drawing of the monitoring and recording system to measure the tissue concentration of fluorescein isothiocyanate (FITC)-dextran in the rabbit tenuissimus muscle

intravenous injection was recorded with a video system and analyzed with a videodensitometer and a computer.

The photographs in Fig. 3 show the reproduced images of microcirculation around a venular end of a capillary at various times after the injection of FITC-Dextran 70. This figure shows that the tracer concentration in the tissue gradually increased from 1 min to 20 min and up to 40 min following the tracer injection. By using these videoimages, the temporal changes in the tracer concentration in the tissue around a single capillary were quantified and analyzed.

FIG. 3. Photographs showing the temporary changes in the tissue concentration of FITC-dextran following its intra-venous injection. The bright portion in the *middle* of each figure indicates a capillary (*right side*) and a consecutive venule (*left side*)

FITC-Dextran (M.W.=70,000)

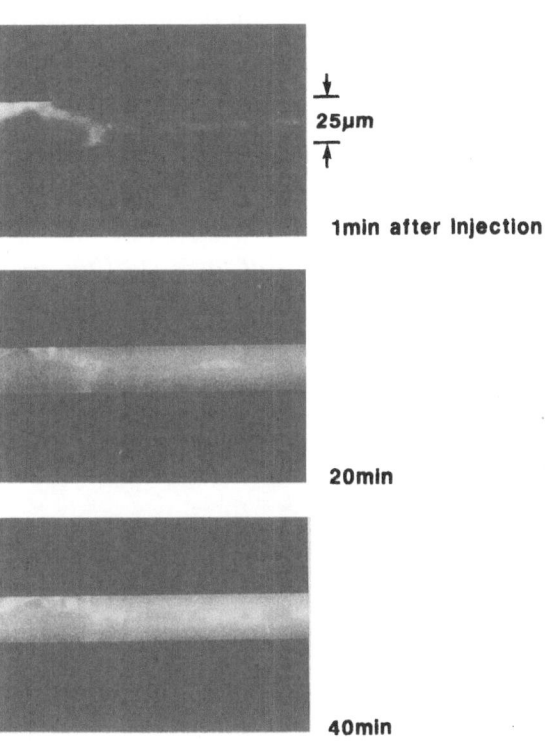

25μm

1min after Injection

20min

40min

A Theoretical Model for Curve Fitting

In order to determine the permeability parameters, the tissue concentration curve against time was fitted with a theoretical curve, which had been derived from the large-pore model. The solute flux through the large pore (Js) can be written as:

$$J_s = -D(dC/dx) + VC \qquad (1)$$

where C is the tracer concentration in the pore at length x from the luminal entrance at time t. D indicates the diffusivity of the tracer in the pore and V designates the velocity of the tracer conveyed by water flux through the pore. The two parameters D and V are expressed with the diffusion permeability (P_d) and water flux (J_v) through the large pore as:

$$D = P_d\Delta X \quad \text{and} \quad V = J_v(1 - \sigma) \qquad (2)$$

where ΔX is the length of the pore and σ is the reflection coefficient of the tracer to the pore. The term $(-DdC/dx)$ in Eq. 1 represents the diffusive component and the term (VC) the convective component of the flux. Solving

Eq. 1 under the steady state condition, by assuming constant concentrations of the tracer in plasma and interstitial fluid (C_p and C_i), the following Patlak's equation [8] is derived with the Peclet number (Pe $= V/P_d$):

$$J_s = V[C_p - C_i \exp(-\text{Pe})]/[1 - \exp(-\text{Pe})] \tag{3}$$

Obviously, it is very difficult to determine the individual values of D and V from the experimental data of the solute flux (J_s) by using this complicated equation. Instead, for the transient phase, Eq. 1 can be rewritten in terms of a simple second-order partial differential equation as:

$$\partial C/\partial t = D\partial^2 C/\partial x^2 - V\partial C/\partial x \tag{4}$$

With boundary conditions appropriately given to the luminal and abluminal ends, it is easy to compute a curve of the tissue concentration as a function of time. Hence, by changing D and V values systematically, it is possible to obtain the best fitting curve to the experimental data and to determine the values of the D and V parameters.

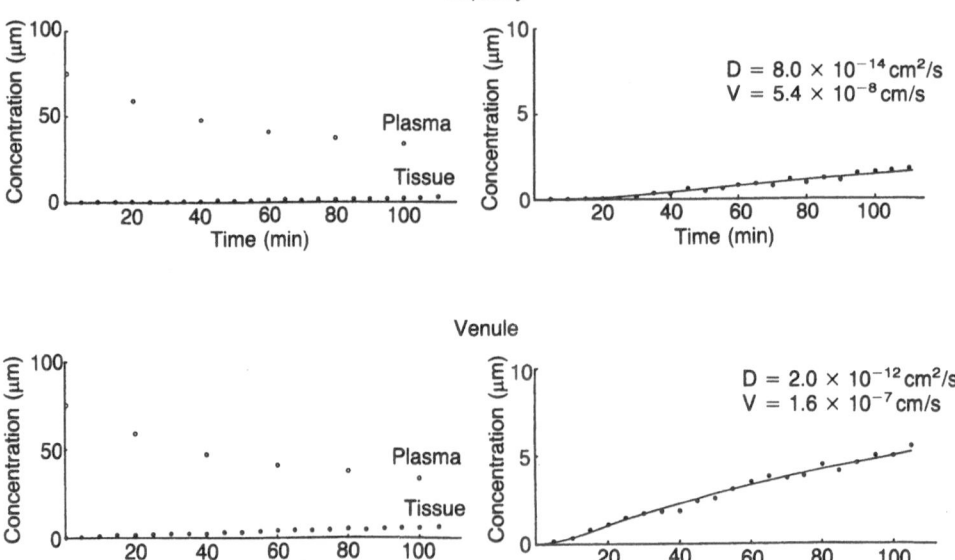

FIG. 4. The time courses of the tracer concentration in plasma and in tissue measured in the same experiment at a capillary (*upper left*) and a venule (*lower left*). The *solid lines* in the *right panels* are the curves of best fit to the measured data, plotted on a larger scale. The diffusivity (D) and velocity (V) values listed in the *right panels*, obtained from the best fitting curves, were used to determine the permeability parameters

Results

Figure 4 depicts typical time courses of the concentration changes of FITC-Dx 70 in the plasma and in the tissue obtained at the true capillary and the venular capillary. The right side of the figure shows the curves of best fit to the data on the left, but plotted on a larger scale. The best fitting curves gave the values of the permeability parameters D and V shown on the top right of each curve.

The values of D and V obtained in the animal experiments are listed in Table 1. Although the data may not be sufficient, it is apparent that in macromolecular transport, the convective component is dominant compared to the diffusive component at any molecular size studied, and that the diffusivity of the tracer at the venular end is approximately 10 times greater than that at the true capillary.

If a set of these D and V data is obtained for a tracer of a certain molecular size, it is possible to estimate the diffusion permeability, the hydraulic conductivity, and the reflection coefficient of the molecule with respect to large pores, as well as the pore size and density at the measured site of a single capillary, by the theoretical treatment described below (cf. [6]).

1. The diffusion permeability (P_d) is calculated from D by:

$$P_d = D/\Delta X \tag{5}$$

where ΔX is the pore length ($= 0.9\,\mu m$).

2. V and the molecular radius (a_e) of the tracer determine the reflection coefficient (σ) and the hydraulic conductivity (Lp) for a given radius of pore (r_p) from the following equations:

$$\alpha = a_e/r_p$$

$$\sigma = (16\alpha^2 - 20\alpha^3 + 7\alpha^4 - \alpha^5)/3$$

$$L_p = J_v/(\Delta p - \sigma\Delta\pi) = V/(1 - \sigma)/(\Delta p - \sigma\Delta\pi) \tag{6}$$

where Δp and $\Delta\pi$ indicate the hydrostatic and osmotic pressure differences across the capillary wall.

3. Pd is theoretically related to L_p and r_p:

$$P_d = [4RT/(3\pi Na)](L_p/r_p^3)(1 - \alpha)^2 F(\alpha)/\alpha, \text{ where}$$

$$F(\alpha) = 1 - 2.104\alpha + 2.09\alpha^3 - 0.95\alpha^5 \tag{7}$$

since,

$$L_p = (A_p/S)r_p^2/(8\mu\Delta X)$$

$$P_d = (A_p/S)(1 - \alpha)^2 F(\alpha)Df/\Delta X$$

$$D_f = RT/(6\pi\mu a_e N_a), \tag{8}$$

where R, T, N_a, μ, and D_f are the gas constant, absolute temperature, Avogadro number, medium viscosity, and free diffusion coefficient,

TABLE 1. Permeability parameters obtained in the rabbit tenuissimus muscle.

Tracer	a_e (nm)	$V(\times 10^{-8} \text{cm/s})$		$D(\times 10^{-12} \text{cm}^2\text{/s})$	
		Capillary	Venule	Capillary	Venule
FITC-Dx 40	4.37	—	22.0 (1)	—	13.0 (1)
FITC-Dx 70	5.79	8.85 ± 2.86 (4)	17.7 ± 3.65 (5)	0.47 ± 0.28 (4)	3.64 ± 2.84 (5)
FITC-Dx 150	8.25	3.40, 5.40 (2)	16.5, 10.0 (2)	0.02, 0.03 (2)	0.50, 0.20 (2)

FITC-Dx, fluorescein isothiocyanate–dextran; A_e, molecular radius; V, velocity; D, diffusivity

respectively. The term (A_p/S) indicates the pore area per unit capillary surface area. Accordingly, it is possible to find an appropriate r_p value which fits the P_d value calculated using Eq. 7 to that estimated from the D value by Eq. 5. Thus, we can derive estimates for r_p, P_d, L_p, and σ.

4. The r_p and L_p values estimated as above allow us to assess the large pore number per unit capillary surface (n/S) as:

$$n/S = 8\mu L_p \Delta X/(\pi r_p^4)$$

by assuming water flux through large pores is laminar.

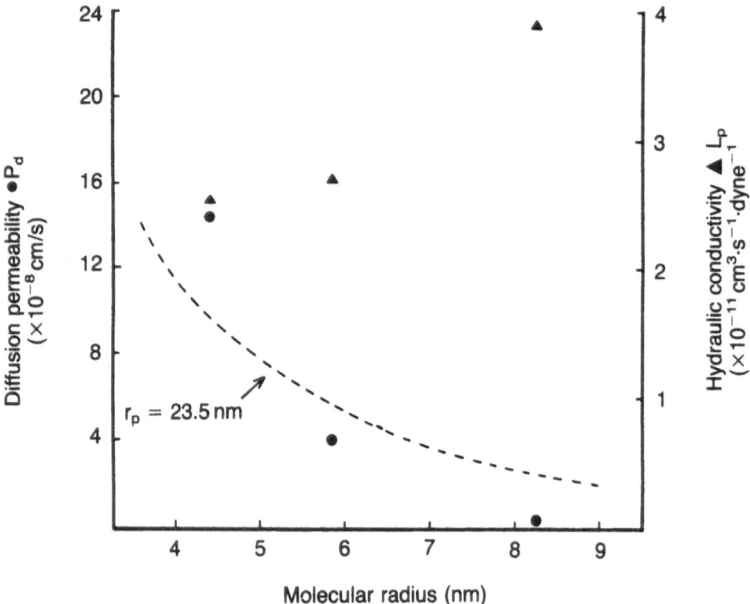

FIG. 5. Estimation of the large pore radius (r_p) using the data of diffusion permeability (P_d) and hydraulic conductivity (L_p) for three FITC-dextrans (FITC-Dx 40, FITC-Dx 70, and FITC-Dx 150) at the venule. The Pd and Lp values are calculated from D and V data listed in Table 1, according to Eqs. 5–7, and are plotted against the molecular radius (a_e) of FITC-Dx 40, FITC-Dx 70 and FITC-Dx 150 $(a_e = 4.37, 5.79, \text{ and } 8.25 \text{ nm},$ respectively). See text for details of the estimation procedures

When a set of D and V values for several tracers of different molecular radius is available, a similar approach can be employed to estimate r_p which gives the best fitting curve to the P_d data by Eq. 5 but no significant difference between the L_p values among the individual tracers. This approach can yield statistically more reliable estimates of r_p, L_p, σ, and n/S. Figure 5 shows an example of such analyses attempted for the data on the venular capillary in Table 1. The dotted line indicates the best fitting curve to the three P_d data points according to an r_p value of 23.5 nm. L_p values estimated from this result are around $(8-10) \times 10^{-12}\,\mathrm{cm}^3\cdot\mathrm{s}^{-1}\cdot\mathrm{dyne}^{-1}$. Since the hydraulic conductivity through small pores is reported to be $25 \times 10^{-12}\,\mathrm{cm}^3\cdot\mathrm{s}^{-1}\cdot\mathrm{dyne}^{-1}$ [4], the conductivity through large pores is found to be approximately 1/3 of that through small pores at the venular capillaries.

Discussion

The results of this study have suggested that the present experimental and theoretical approaches enable us to have quantitative information about the morphological parameters of large pores in a single capillary wall (pore size and density), together with the related membrane parameters (the diffusion, permeability, and reflection coefficient of macromolecules and the hydraulic · conductivity of water flux through the pores). If D and V data measured at different sites along the capillary channel were available, we could visualize the distributions of these parameters along the channel.

The analysis in this study has been carried out by assuming the existence of large pores. The validity of this concept should be carefully verified with some experimental and/or theoretical approaches. However, the values of large pore radius (r_p) and the reflection coefficients (σ) estimated in this study in the skeletal muscle microvessels agree well with those in the lung, heart, skin, etc., obtained by pore-stripping analyses [9] based on data on the lymph/plasma concentration ratio of various tracers in these organs. This agreement may be regarded as evidence supporting the validity of the present attempt, since all these organs are known to have the continuous capillary in common. It is, however, difficult to compare the values of diffusion permeability (P_d) and of hydraulic conductivity (L_p) estimated here with the data reported in the literature. This is because the conventional permeability values estimated by whole organ measurement (e.g., [10]) include the substantial component caused by the double-pore effect [5] and do not represent the novel diffusion permeability through large pores. With respect to L_p, to date there has been no report which has quantified the conductivity of water flux across large pores. The values of L_p and σ or more directly V values estimated in this study have made it possible first to calculate the steady-state solute flux of macromolecules across the capillary wall by Patlack's equation (see Eq. 3).

The results in this study show that the estimated value of L_p through large pores at the venular capillary is nearly 1/3 of L_p through small pores. If all the assumptions employed in this study are valid, this result implies that a fairly large flux of water may occur through large pores. In addition, the direction of this flux must always be from blood to tissue, since the reflection coefficients of plasma proteins do not seem large enough to cause effective osmosis. Hence, this outward fluid flux through large pores may substantially affect the water balance along the capillary channel, which has conventionally been explained, for small pores, by a scheme invoking an ultramicrocirculation [11].

References

1. Hudlicka O (1984) Development of microcirculation: capillary growth and adaptation. In: Handbook of physiology, Sec 2, Vol 4. American Physiological Society, Bethesda, pp 165–216
2. Harlan JM (1985) Leukocyte-endothelial interactions. Blood 65:513–526
3. Becker AY, Ritter AB, Duran WN (1982) Analysis of microvascular permeability to macromolecules by video-image digital processing. Microvasc Res 38:200–216
4. Renkin EM (1985) Capillary transport of macromolecules: pores and other endothelial pathways. J Appl Physiol 58:315–325
5. Rippe B, Kamiya A, Folkow A (1979) Transcapillary passage of albumin, effect of tissue cooling and of increases in filtration and of plasma colloid osmotic pressure. Acta Physiol Scand 105:171–187
6. Curry FE (1984) Mechanics and thermodynamics of transcapillary exchange. In: Handbook of physiology, Sec 2, Vol 4. American Physiological Society, Bethesda, pp 309–374
7. Bass L, Robinson PJ (1982) Capillary permeability of heterogenous organs: a comparisoneous interpretation of indicator diffusion data. Clin Exp Pharmacol Physiol 9:363–388
8. Patlack CS Rapoport SI (1971) Theoretical analysis of net tracer flux due to volume circulation in a membrane with pores of different sizes. Relation to a solute drag model. J Gen Physiol 57:113–124
9. Taylor AE, Granger DN (1984) Exchange of macromolecules across the microcirculation. Handbook of physiology, Sec 2, Vol 4. American Physiological Society, Bethesda, pp 467–520
10. Garlick DG, Renkin EM (1970) Transport of large molecules from plasma to interstitial fluid and lymph in dogs. Am J Physiol 219:1595–1605
11. Landis EM, Pappenheimer JR (1963) Exchange of substances through capillary wall. In: Handbook of physiology, Sec 2. American Physiological Society, Bethesda, pp 961–1034

Physiology and Functional Anatomy of the Venous System

Toshio Ohhashi[1], Keiko Morimoto-Murase[2], and Takeshi Kitoh[1]

Abstract. To study regional differences of mechanical properties of veins and the contribution of tissue components to these properties, pressure-volume relationships were obtained with cylindrical segments of isolated dog peripheral and trunk veins. The transmural pressure of the segment was raised up to $20\,cmH_2O$ and then reduced to $0\,cmH_2O$ by increasing and decreasing the intraluminal volume at a constant rate. The pressure volume relationships of venous segments were also constructed in the same way after treatment with $1\,mg/ml$ collagenase for $30\,min$, $0.1\,mg/ml$ elastase for $5\,min$, or $1\,mg/ml$ hyaluronidase for $60\,min$. The pressure-volume curves obtained in the veins of lower and upper extremities and the jugular veins are curvilinear with considerable convexity toward the volume axis. It is apparent that most of the volume increment results from tangential distension within the physiological range of venous pressure. On the other hand, the pressure-volume curves in the intrathoracic and suprarenal inferior caval veins were concave at lower pressure range and convex at higher pressure toward the volume axis. A transmural pressure producing collapse of the caval veins was about $-8\,cmH_2O$, lowest in all veins. There were marked regional differences in mechanical properties of the veins. The non-uniform distensibility of the venous walls was related to local differences in architecture of collagen and elastic fiber components and the content of smooth muscles in the walls. Treatment with collagenase or elastase produced a significant increase of the incremental volume elasticity within the pressure range of $0–2\,cmH_2O$ (E_{0-2}). The treatment, however, caused no effect on E_{10-20}. Treatment with hyaluronidase induced no effect on these mechanical parameters. The anatomical and functional characteristics of the distribution of venous valves were also examined. In the venous trees of humans, monkeys, dogs, and rabbits, the valves with bicusps were always found at the positions of the venous angles and under the inguinal ligaments, but not in the superior and inferior caval veins. When a valve exists,

---

[1] The First Department of Physiology, Shinshu University School of Medicine, 3-1-1 Asahi, Matsumoto, 390 Japan
[2] Present address: Faculty of Art, Yamaguchi University, Yamaguchi, 753 Japan

a confluence of tributaries is always able to find in the upstream compartment of the main vein. The valves are always found at positions where the veins are compressed mechanically by movement of the joints, or stretch of the tendons or fasciae.

Key words: Pressure-volume relationship—Incremental volume elasticity— Regional differences—Collagenase—Elastase—Venous valves—Law of distribution of venous valves

Introduction

The main physiological role of the veins is to adjust the capacity of the vascular system, ensuring an appropriate return of blood to the heart. The veins are also a determinant of capillary pressure, ensuring a favorable internal environment for cells of the body.

The veins can behave as a passive reservoir, due to the tissue components of their wall and the presence of valves in the system. The amount of blood which is passively mobilized from this reservoir towards the heart is determined by the venous distending pressure, which is dependent on the hydrostatic pressure load and the degree of arteriolar constriction. The veins contain adrenergically innervated smooth muscles. The splanchnic capacitance vessels can act as an active blood mobilization system under the sympathetic control [1–3].

Despite these important aspects, surprisingly little is known about regional differences in the visco-elastic properties of isolated veins, the contribution of the wall components to the mechanical properties, or the anatomical principles in the distribution of venous valves. The purpose of this chapter is to introduce some of our recent studies concerning physiological characteristics and visco-elastic properties of the venous system, and functional anatomy of the venous valves.

Regional Differences in the Venous Distensibility

Thirty-nine mongrel dogs of both sexes, weighing 10–20 kg, were anesthetized with sodium pentobarbital (25–30 mg/kg, i.v.) and killed by bleeding. After marking lengths of exactly 2 cm or 3 cm in vivo, cylindrical segments of the external jugular, axillar, brachial, external iliac, femoral, saphenous, and superior and inferior caval veins without branches, were quickly dissected and then cleaned of adipose and connective tissues. Each end of the venous segment was cannulated with a methylacrylate tube. To ensure that the diameter was uniform along the length of the segment, care was always taken to match closely the size of the tubing to the diameter of the segment at 0 cmH₂O of transmural pressure. The segment was stretched to its in vivo length and placed in an organ bath filled with Krebs-bicarbonate solution circulating through a heat exchanger kept at 37°C. The composition of the solution was as follows (in mM): 120.0 NaCl, 5.9 KCl, 25.0 NaHCO$_3$, 1.2 NaHPO$_4$, 2.5 CaCl$_2$, and 5.5 glucose. The solution was equilibrated before and during the experiment with a gas mixture of 95% O$_2$—5% CO$_2$ in the organ bath to give a pH of 7.4. The inlet cannula led via a three-way stopcock either to a microinfusion pump (Harvard Model 901D) or a Mariotte's bottle. The pump was able to perfuse the Krebs solution into the intraluminal space of the venous segment at a constant flow rate (Q = 0.08 ml/min) so that the instantaneous volume of the segment (V) at any instant of time (t) was V = V$_0$ + (Q × t) if the outlet cannula was connected directly to a pressure transducer (blind system). An electrical integrator started simultaneously with the calibrated delivery rate of the perfusion pump and provided a ramp whose slope was equal to the rate of injection. The voltage at the output of the integrator was proportional to the volume of venous segment. The reference volume (V$_0$) was calculated by using the diameter and length of the venous segment at 0 cmH$_2$O of the transmural pressure. The hydrostatic head of pressure around the venous segment was adjusted to 1 cmH$_2$O during measurement by keeping the height of Krebs solution above the specimen at constant level. The diameter of the venous segment was continuously measured with a domestic-made noncontact diameter gauge utilizing image sensor [4]. By use of the two diameter gauges, changes in the diameters at right angles were measured to confirm that the cross-sectional area was nearly circular at zero transmural pressure. The transmural pressure of the venous segment was measured with the low pressure transducer (Toyo-Baldwin 0.1-350-0) which was placed at the same height as that of the venous segment via a three-way stopcock connected with the outlet cannula.

All venous segments were allowed to equilibrate for 30 min with intraluminal perfusion of the 37°C Krebs solution at a constant transmural pressure (about 1 cmH$_2$O) through a Mariotte's bottle, before the start of experiment. After turning the lever of the stopcock connected with the outlet cannula to make up the blind system and then adjusting the transmural pressure to 0 cmH$_2$O, pressure-volume relationships for the venous segments were obtained by increasing and decreasing the intraluminal volume of the segment at a constant rate (0.08 ml/min) with the micro-infusion pump. In each cycle, the transmural pressure of the segment was raised to 20 cmH$_2$O and then reduced to 0 cmH$_2$O. During the period of pressure increase, no bulging of the venous segment was observed at any point along its length. After preconditioning with several inflation-deflation cycles, two reproducible pressure-volume curves were recorded on an X-Y recorder (Watanabe, WX 4402).

To investigate effects of activation of venous smooth muscle on the mechanical properties of the isolated veins, we experimented with a modified Krebs solution containing a supramaximal dose of norepinephrine hydrochloride (10^{-4} M). The solution was perfused intraluminally for 30 min after the control pressure-volume relationship was obtained with each venous segment. Upon activation of the smooth muscle, pressure-volume relationships were quickly

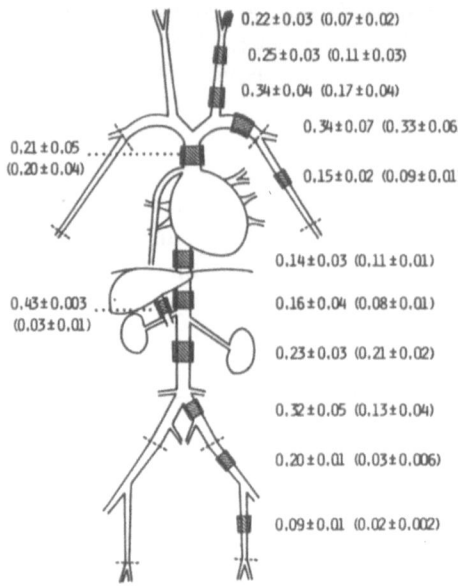

0.22±0.03 (0.07±0.02)

0.25±0.03 (0.11±0.03)

0.34±0.04 (0.17±0.04)

0.34±0.07 (0.33±0.06)

0.21±0.05
(0.20±0.04)

0.15±0.02 (0.09±0.01)

0.14±0.03 (0.11±0.01)

0.43±0.003
(0.03±0.01)

0.16±0.04 (0.08±0.01)

0.23±0.03 (0.21±0.02)

0.32±0.05 (0.13±0.04)

0.20±0.01 (0.03±0.006)

0.09±0.01 (0.02±0.002)

FIG. 1. The incremental volume elasticity at transmural pressures ranging from 0 to 5 cmH$_2$O (E_{0-5}) with (*intraparenthesis*, $n = 4$) and without (*extraparenthesis*, $n = 4$) activation of venous smooth muscle, in the isolated dog trunk and limb veins. (From [3] with permission)

generated, just after assuring that the venous tissues were in a steady state of response to the test solution.

After the end of these experiments, the venous segments were fixed at their in vivo length with a 10% formalin solution, and the sections were stained with Mallory-Azan or the elastica Van-Gieson technique.

Quasi-static properties of the venous segments were compared by use of an incremental volume elasticity (wall distensibility) at transmural pressures ranging from 0 to 5 cmH$_2$O (E_{0-5}). Volume elasticity can be calculated as $\Delta V \times V_0^{-1} \times \Delta P^{-1}$ (where ΔV, V_0, and ΔP denote incremental and reference volumes of the venous segments, and incremental transmural pressure), i.e., the fractional change in volume of the venous segment due to the 5 cmH$_2$O increment in the transmural pressure.

Figure 1 summarizes the average values of wall distensibility (E_{0-5}) with or without the activation of venous smooth muscle, calculated from the pressure-volume relationships in the isolated dog trunk and limb veins. The numbers intra- and extra-parentheses show the average and standard error of each wall distensibility (E_{0-5}) with and without the activation of venous smooth muscle, respectively. As shown in this figure, there are marked regional differences in the incremental volume elasticities (E_{0-5}) with and without the activation of smooth muscle. Of all the isolated dog veins, the greatest venous wall distensibility was observed in the hilar portal vein. Activation of the venous smooth muscle in the portal vein also caused the largest reduction of the distensibility (E_{0-5}) in all the veins tested.

Figure 2 shows representative recordings of pressure-volume relationships before and after treatment with $10^{-4} M$ norepinephrine (NA) in the isolated

FIG. 2. Representative recordings of pressure-volume relationships of the isolated dog hilar portal vein before (*control*) and after (*$10^{-4} M$ NA*) activation of the venous smooth muscle induced by $10^{-4} M$ norepinephrine (NA). (From [3] with permission)

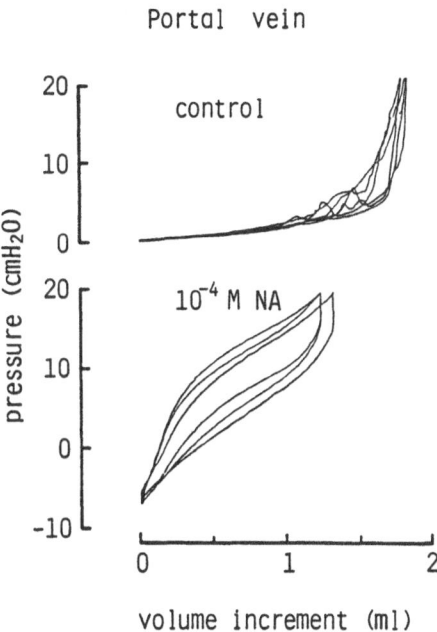

Portal vein

dog extrahepatic hilar portal vein. As shown in the upper panel of Fig. 2, the pressure-volume curves obtained at the in vivo length are curvilinear with considerable convexity toward the volume axis. It is apparent that most of the volume increment results from tangential distension within the physiological range of venous pressure. Thus, the relationship was nearly linear between 0 and $2-3 \, cmH_2O$ of the transmural pressure. It is only with pressures of more than $10 \, cmH_2O$ that tangential distension becomes restricted to the reactively low degree of distensibility. The increasing volume of the venous segments also caused an appearance of spontaneous contractions of the venous smooth muscles, which produced clearly phasic undulations on the pressure-volume curves.

The lower panel of Fig. 2 demonstrates representative recordings of the pressure-volume relationships of the same venous segment after activation of the venous smooth muscle induced by $10^{-4} M$ NA. During the inflation phase, the volume changed only slightly at the lower pressure range ($0-5 \, cmH_2O$) and then to a greater extent at pressures higher than $5 \, cmH_2O$. During the activation of venous smooth muscle, the incremental volume elasticity was smaller than that during rest, over part of the physiological pressure range ($0-5 \, cmH_2O$, E_{0-5}), and greater than during rest over other parts of the range ($10-20 \, cmH_2O$). The curves during the activation also showed a larger hysteresis than that observed in the control. The E_{0-5} calculated with the venous segments during the activation was $0.03 \pm 0.01 \, cmH_2O^{-1}$ ($n = 4$).

The wall thickness of the portal vein was greatest in the dog trunk and limb veins used, being approximately similar to that of suprarenal inferior caval vein. Smooth muscles of the portal vein are divided into an outer longitudinal layer and an inner layer arranged as a low-pitched helix. The longitudinal layer contains many mitochondria and pinocytotic vesicles, which suggests high levels of metabolic activity. The circular layer is thinner, has sparse innervation, and appears to show lower levels of metabolic activity [5]. Unmyelinated nerves traverse the outer longitudinal layer and form a dense plexus between the outer and inner layers [5]. The media of the portal vein also consists of lots of collagen fibers and a few elastic fibers, which are localized among the circular and longitudinal smooth muscle layers. The structure of the portal vein is also similar to that of the inferior caval vein (see Fig. 4). The morphological similarity may be correlated with the evidence that the portions of the trunk veins are developed from the same parts of the embryological trunks [6].

Besides the smooth muscle cell arrangement in the walls of the veins, there is also a varying amount of elastin and collagen, which confer on the vessel its passive and active mechanical properties. It would appear likely that the smooth muscle cells not only subserve the function of contracting and changing

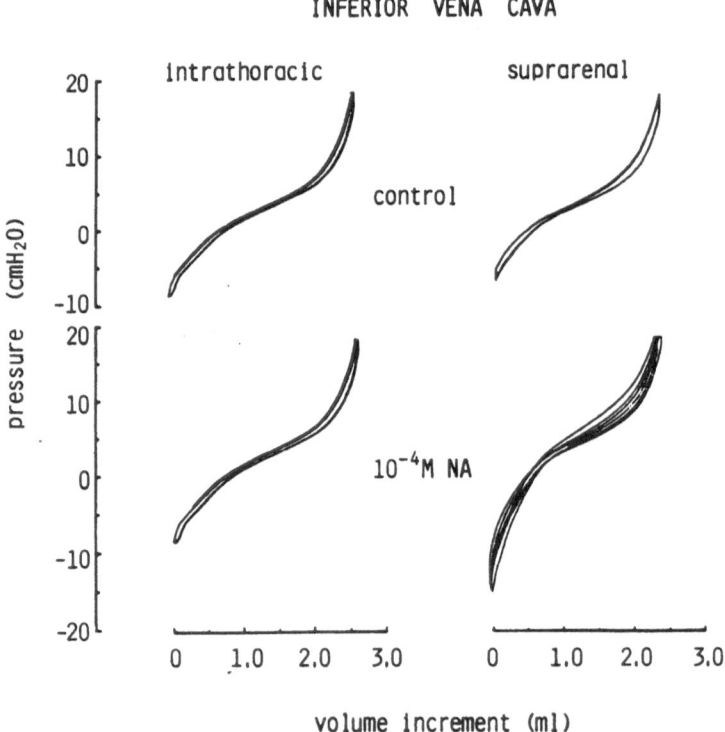

FIG. 3. Representative recordings of pressure-volume relationships before (*control*) and after ($10^{-4} M NA$) activation of venous smooth muscle by $10^{-4} M$ norepinephrine (NA) in isolated dog inferior caval veins at the intrathoracic and suprarenal portions. (From [3] with permission)

the tone in the vessel wall, but they are also capable of synthesizing collagen and elastin, and this, as Somlyo and Somlyo [7] point out, gives the smooth muscle cell an importance in the generation of diseases in which the visco-elastic elements of the venous wall may be abnormal [8].

As shown in Fig. 1, in the cases of trunk veins, wall distensibility (E_{0-5}) in the inferior caval veins passing through the diaphragm was significantly lower than that in the superior and infrarenal inferior caval veins.

Figure 3 shows representative recordings of pressure-volume relationships before and after activation of venous smooth muscle by $10^{-4} M$ NA, in dog inferior caval veins isolated from the intrathoracic and suprarenal portions. As shown in the upper panels of Fig. 3, the pressure-volume curves before the activation of the venous smooth muscle were nonlinear and concave at lower pressures, and convex toward the volume axis at higher pressures, indicating an inflection point at about $2 \, cmH_2O$. The transmural pressure inducing collapse of the venous segments in the intrathoracic and suprarenal inferior caval veins was about $-8 \, cmH_2O$, the value being significantly lower than those in the superior and infrarenal inferior caval veins (about $-2 \, cmH_2O$). As shown in the lower panel, the activation of venous smooth muscle in the suprarenal inferior caval vein caused a significant decrease of the transmural pressure needed to induce the collapse, and no change of the transmural pressure at the inflection point of the pressure-volume curves.

Superior vena cava

Inferior vena cava (supradiaphragma)

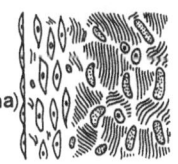

FIG. 4. Schematic illustrations of transverse sections of the venous walls in the dog superior and inferior caval veins. Each section shows an averaged wall thickness of the veins. The spindle-like cells in the *left side* and media layer of the sections are endothelial and smooth muscle cells, respectively. The areas represented by *dotted* and *wavy lines* denote the elastic and collagen fibers, respectively

Inferior vena cava (between liver and renal veins)

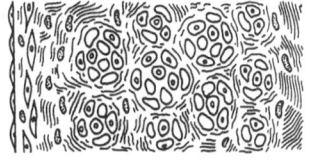

Inferior vena cava (infrarenal)

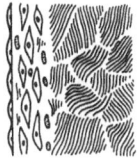

On the other hand, the activation does not affect significantly the shape of the pressure-volume curve or the transmural pressure inducing the collapse in the intrathoracic inferior caval vein. The mechanical properties of the intrathoracic and suprarenal inferior caval veins passing through the diaphragm may be related to a preventive function against venous collapse. If this were not the case, rhythmical contractions of striated muscles in the diaphragm would produce collapse of the veins, inhibition of venous return, and then syncope.

Figure 4 is a schematic illustration of transverse sections of venous walls in the dog superior and inferior caval veins. In the superior and inferior caval veins, the venous wall consists of two circumferential smooth muscle layers and a large number of collagen fibers. The wall thickness of the veins is about 0.4 mm. On the other hand, the venous wall in the intrathoracic inferior caval vein has more than three smooth muscle layers in the media and a small number of elastic fibers in the media and adventitia. The collagen fibers in the wall are longitudinally overstretched at its in vivo length. In contrast to the veins mentioned previously, in the suprarenal inferior caval vein the wall was about 2.5 times as thick as those in the other trunk veins. Smooth muscle cells in the media are well developed and arranged into two layers: the internal circumferential and external longitudinal. These structural features may be a morphological reflection of the observed lower distensibility in the inferior caval veins passing through the diaphragm.

Contribution of Wall Components to Venous Distensibility

Thirty mongrel dogs of both sexes, weighing 7–20 kg, were anesthetized with sodium pentobarbital (30 mg/kg, i.v.) and killed by bleeding. After marking lengths of exactly 3 cm in vivo, cylindrical segments of the external jugular veins without branches were quickly dissected and then cleaned of adipose and connective tissues. The experimental procedure and protocol to obtain the pressure-volume relationships of the isolated veins have been described previously [3]. In order to evaluate the contribution of wall components to the quasi-static properties, the venous segments with the cannulas were removed from the organ bath after control pressure-volume relationships had been obtained at their in vivo length. The segments were then incubated at 37°C in oxygenated Krebs solution containing one of the following enzymes: collagenase type I (315 units/mg protein; 1 mg/ml) for 30 min; elastase (340 units/mg protein; 0.1 mg/ml) for 5 min; or bovine testis hyaluronidase (295 units/mg protein; 1 mg/ml) for 60 min. Segments were then returned into the organ bath in the same position as that in the control experiment. Pressure-volume relationships of the enzyme-treated venous segments were constructed 30 min after intraluminal perfusion of the enzyme-free Krebs solution [9].

Quasi-static properties of the venous segments were compared by use of two kinds of incremental volume elasticities, at transmural pressures ranging from 0 to $2 \text{cmH}_2\text{O}$ (E_{0-2}) and 10 to $20 \text{cmH}_2\text{O}$ (E_{10-20}). Volume elasticity can be calculated as $\Delta V \times V_0^{-1} \times \Delta P^{-1}$, where ΔV, V_0, and ΔP denote incremental and reference volumes of the venous segments, and incremental transmural pressure.

After the end of these experiments, histological studies were performed in the same way as that described previously [3].

The pressure-volume curves in the isolated dog external jugular veins before the enzymatic treatment were curvilinear with considerable convexity toward the volume axis. Thus, the curve was nearly linear between 0 and $2-3 \text{cmH}_2\text{O}$ of transmural pressure. Treatment with collagenase caused a moderate decrease in slope over the linear part of the curves at the lower pressure range, which resulted in a greater increment of the volume within the pressure range compared with that in the control ($0.26 \pm 0.01 \text{cmH}_2\text{O}^{-1}$ on treatment with collagenase, vs $0.22 \pm 0.01 \text{cmH}_2\text{O}^{-1}$ in the control, $P < 0.05$, $n = 4$). No significant difference in the E_{10-20}, however, was observed between the venous segments before [$(4.4 \pm 0.4) \times 10^{-3} \text{cmH}_2\text{O}^{-1}$] and after [$(4.3 \pm 0.5) \times 10^{-3} \text{cmH}_2\text{O}^{-1}$] treatment with collagenase.

Similar findings were observed with the venous segments before and after treatment with 0.1mg/ml elastase for 5 min. The slope in the linear part of the pressure-volume relationships at lower pressure range was slightly reduced by the enzymatic treatment. Thus, the E_{0-2} in isolated external jugular veins was significantly increased by treatment with elastase ($0.26 \pm 0.01 \text{cmH}_2\text{O}^{-1}$ on treatment with elastase, vs $0.22 \pm 0.01 \text{cmH}_2\text{O}^{-1}$ in the control, $P < 0.05$, $n = 4$). On the other hand, no significant difference in the E_{10-20} was observed before [$(4.5 \pm 0.7) \times 10^{-3} \text{cmH}_2\text{O}^{-1}$] or after treatment with elastase [$(4.0 \pm 0.5) \times 10^{-3} \text{cmH}_2\text{O}$].

Treatment with 1mg/ml hyaluronidase for 60 min did not significantly affect the E_{0-2} ($0.23 \pm 0.01 \text{cmH}_2\text{O}^{-1}$ on treatment with hyaluronidase, vs $0.23 \pm 0.01 \text{cmH}_2\text{O}$ in the control, not significant, $n = 4$), or the E_{10-20} [$(4.5 \pm 0.1) \times 10^{-3} \text{cmH}_2\text{O}^{-1}$ on treatment with hyaluronidase, vs $(4.3 \pm 0.3) \times 10^{-3} \text{cmH}_2\text{O}^{-1}$ in the control, not significant, $n = 4$].

Histological studies demonstrate that the intima of the dog external jugular veins consists of endothelial cells. Smooth muscle cells are present in the vein, closely apposed to the endothelial cells. The media of the veins has several smooth muscle layers connected to elastic and collagen fibers. A felty network of collagen bundles was also observed throughout the wall, with interwoven elastic and muscular fibers in the media.

Histological sections of venous segments fixed after treatment with elastase revealed that the enzyme caused fragmentation of elastic fibers in the media and adventitia, which was also confirmed with a general decrease in uptake of elastica Van-Gieson stain. The fragmentation produced a breakage of the network of collagen fibers in the adventitia and then a swelling of the venous wall. In contrast, the felty network of collagen fibers in the media and adventitia seemed to be disrupted by treatment with collagenase, but elastic fibers in the media and adventitia, stained yellow in a bead-like shape, remained as almost the same as those in the control. Venous sections fixed after treatment with hyaluronidase showed no significant change compared with controls.

It is well known that the mechanical properties of blood vessels are determined by the relative amounts of the various wall components, their individual physical properties, and the way in which they are architecturally coupled. There have been numerous studies to investigate the contribution of the various components to the mechanical properties of arteries and aortas [10–12]. Little information, however, has been reported on the contribution of the components to the mechanical properties of veins. In the present study, we focused on demonstrating clearly the contribution of wall components to the quasi-static properties of isolated dog external jugular veins. The present results show that treatment with collagenase or elastase causes a significant increase of the E_{0-2} of the venous segments. No appreciable change in the E_{10-20} value was observed following treatment with collagenase or elastase. Disruption of the felty network of collagen bundles in the media and adventitia, produced by collagenase treatment, may be responsible, in part, for the significant increase of the E_{0-2}. It is also likely that elastase did, in fact, disrupt or even destroy the integrity of elastin, because elastase produced both histological and mechanical changes in the veins.

On the other hand, treatment with 1 mg/ml hyaluronidase for 60 min had no significant effect on the E_{0-2} and E_{10-20} at the vein's in vivo length, but it caused a significant attenuation of the longitudinal extension-induced decrease of the E_{0-2} and the extension-induced increase of the E_{10-20} [9]. These results suggest that the glycosaminoglycans (GAGs) located in the venous wall may contribute, in part, to the venous distensibility in the longitudinal direction but not in the circumferential. This may be related to anisotropic differences in the distribution of the GAGs in the venous wall, in the interaction of GAGs with collagen and elastin, or in the chemical characteristics of the GAGs.

Functional Anatomy of Venous Valves

Unlike the arterial system, the venous system has valves that normally prevent retrograde flow [13]. The competence of these valves is crucial for the maintenance of reasonably low capillary pressures of the dependent limbs of humans, because they are an essential part of the muscle-pump mechanism that aids venous return from the legs while walking or moving [14]. This pumping action can reduce the hydrostatic pressure from the ankle to the heart during walking to about one-fourth of that without limb movement [15].

Figure 5 shows a representative diagram of the distribution of venous valves of bicuspid type in a male human being of age 44 who died from stomach cancer [3]. No venous valves were observed in the superior and inferior caval veins, or in the external and internal jugular veins, which are not subjected to an elevated hydrostatic pressure under physiological conditions. Another type

FIG. 5. A representative diagram of the distribution of venous valves of the bicuspid type in a male human being of age 44. (From [3] with permission)

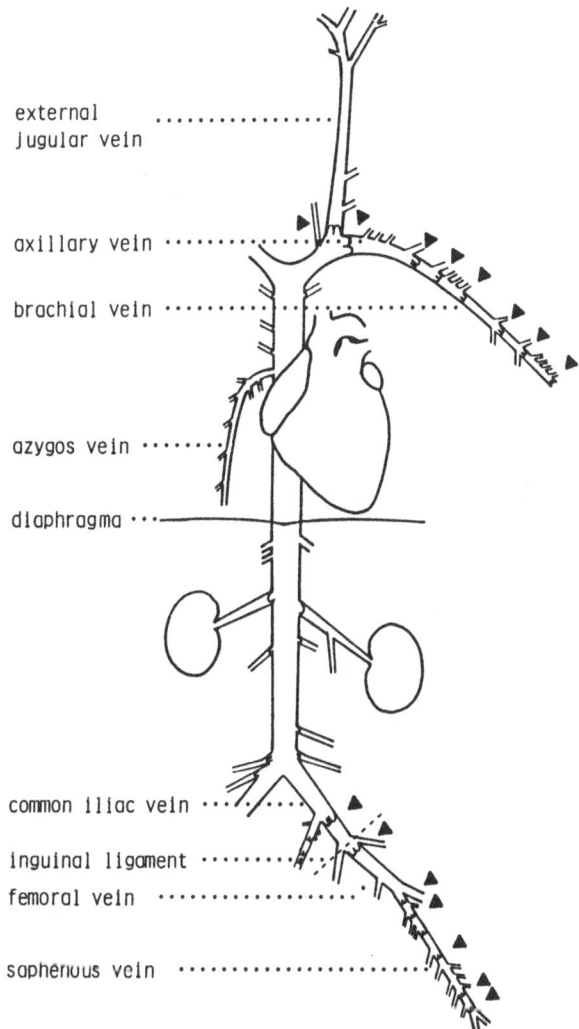

external jugular vein

axillary vein

brachial vein

azygos vein

diaphragma

common iliac vein

inguinal ligament

femoral vein

saphenous vein

of venous valve with a single cusp like a flap, however, was found in the orifices of renal veins connected with the inferior caval veins. These valves may not preclude backflow of blood into the renal veins. The physiological significance of the existence of the flap-like valve is still unknown.

Valves are absent from the several venous plexus in the vertebral venous system, from the cerebral sinuses, and from the cerebral, pulmonary, portal, uterine, and ovarian veins [16]. The hepatic veins do not contain valves, although at the entrance of the hepatic vein into the inferior caval vein there is a semilunar fold, a "sluice valve," which is said to prevent a reflux of blood from the inferior caval vein into the liver. These valves are activated by a relatively thick mass of longitudinal smooth muscle in the wall of the inferior caval vein, contraction of which elevates the fold and the inferior post of the orifice, and narrows the orifice. However, this sluice valve, which is prominent in some animals, is poorly developed in man [17].

The bicuspid venous valves were found at the position of the venous angles and the inguinal ligaments in all of the 12 human beings examined. Valves were often observed at the junction of the external and internal iliac veins, except in people over age 70. Thus, the valve may be rudimentary in older people. It is well known that the common iliac veins have no valves and 75% of external iliac veins have no valves, but only 25% of common femoral veins are valveless. It has been suggested that a lack of valves in the external iliac and femoral veins is the starting point for the development of a progressive descending valvular incompetence that causes varicose veins [18].

Figure 6 shows representative diagrams of the distribution of bicuspid venous valves in the limb venous system in a male human being aged 56 who died from stomach cancer. As shown in Fig. 6, it may be a general tendency that there is a higher density of venous valves in the deep veins than in the superficial veins. This tendency seems to be more remarkable in the lower extremities. Thus, valves are particularly well developed in regions subject to a higher hydrostatic pressure due to gravity. It is also likely that the venous valves are always located distal to the junction of tributaries with the main vein. The venous valves consist of a layer of collagen covered by endothelium. There is usually an area of smooth muscle in the region of the attachment of the valves to the vein wall. The histological findings are compatible with several studies reported previously [8].

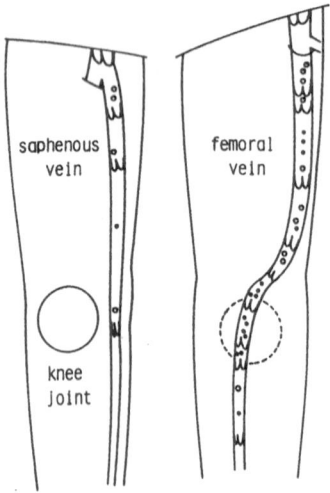
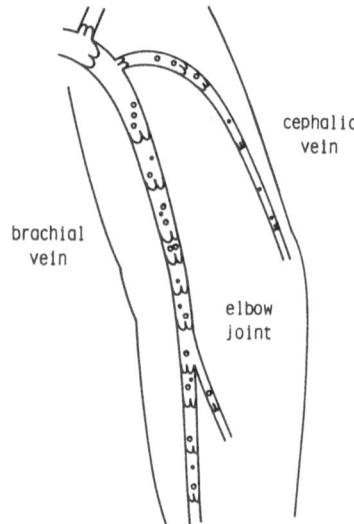

FIG. 6. Representative diagrams of the distribution of valves in limb veins of human upper and lower extremities (subject aged 56)

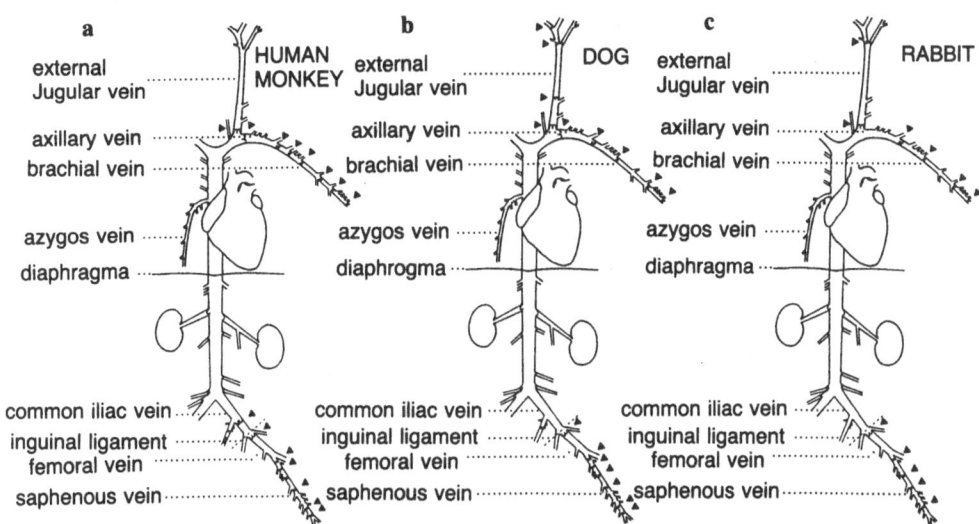

FIG. 7a–c. Representative diagrams of the distribution of venous valves of the bicuspid type in human beings (a), monkeys (a), dogs (b), and rabbits (c). (From [3] with permission)

Another important aspect is the anatomical finding that the venous valves are always found at the regions where the deep limb veins are compressed anatomically or functionally by extrinsic forces such as the movement of joints or the overstretched tension of tendons and fasciae piled up on the veins (Fig. 6). These findings suggest that the venous valves work as a specialized apparatus to prevent not only retrograde blood flow through the valve but also venous collapse or overdistension under physiological conditions.

We also examined from the hydrostatic point of view whether or not there are species differences in the localization of venous valves among human beings, monkeys, dogs, and rabbits. Figure 7 shows representative diagrams of the distribution of venous valves in trunk and deep limb veins of humans, monkeys, dogs, and rabbits. In all animals, there is little doubt that the bicuspid venous valves are always found at the region of the venous angles and the inguinal ligaments, but never observed in the superior or inferior caval veins. Thus, the two valves in trunk veins may play an important role in the regulation of venous return; they are able to separate the reservoir of venous blood into two main compartments. Changes in the wall distensibility of one compartment, constituted by the trunk veins, seem to regulate quickly the venous return. Changes in wall distensibility of a second compartment, composed of the portal, cerebral, and limb veins, may play an important role in slow and precise adjustment of the venous return.

On the other hand, there are marked species differences in the distribution of venous valves in the jugular and limb veins. Thus, no valves are observed in the jugular veins of humans or monkeys, but a great number of the venous valves are found in the external jugular veins of dogs and monkeys. The density of venous valves in the deep limb veins of legs is significantly higher in humans and monkeys than in dogs or rabbits. Furthermore, the number of the venous valves in the deep veins seem to be greater in dogs than in rabbits [3].

In conclusion, functional and anatomical principles in the distribution of venous valves are summarized as follows:

1. In human beings, monkeys, dogs, and rabbits, bicuspid venous valves are always observed at the regions of the venous angles and the inguinal ligaments.
2. In all the animals, the numbers of venous valves in deep limb veins are greater than those in the superficial limb veins.
3. When a venous valve exists in limb veins, a confluence of tributaries in always found in the up-stream compartment of the main vein.
4. The venous valves in deep limb veins are always found at the regions where the veins are compressed by extrinsic forces produced by the movement of joints or by the mechanical compression of tendons and fasciae piled up on the veins.

The physiological role of venous valves may be related to preventing not only retrograde blood flow through the valve but also collapse or overdistension of the vein when the vein is compressed, by flection of the joint, or by mechanical compression of the overstretched tendons and fasciae.

References

1. Vanhoutte PM, Janssens WJ (1978) Local control of venous function. Microvasc Res 16:196–214
2. Rothe CF (1983) Venous system: Physiology of the capacitance vessels. In: Shepherd JT, Abboud FM, Geiger SR (eds) Handbook of physiology, Sect 2: The cardio-vascular system, Vol III: Peripheral circulation and organ blood flow, Part 1. American Physiological Society, Bethesda, pp 397–452
3. Ohhashi T, Morimoto-Murase K (1987) Functional and morphological characteristics of canine veins and venous valves. In: Sakaguchi S (ed) Advances in phlebology, John Libbey, London, pp 34–43
4. Sakaguchi M, Ohhashi T, Azuma T (1979) A photoelectric diameter gauge utilizing the image sensor. Pflügers Arch 378:263–268
5. Brown BP, Anuras S, Heistad PD (1982) Responsiveness of longitudinal and circular muscle layers of the portal vein. Am J Physiol 242:G498–G503
6. Browse NL, Burnand KG, Thomas ML (1988) Diseases of the veins: pathology, diagnosis and treatment. Edward Arnold, London, pp 23–51
7. Somlyo AP, Somlyo AV (1968) Vascular smooth muscle. Pharmacol Rev 20:197–272
8. Cooper KE (1981) Functional aspects of the venous system. In: Schwartz CJ, Werthessen NT, Wolf S (eds) Structure and function of the circulation, Vol 2. Plenum, New York, pp 457–485

9. Kitoh T, Kawai Y, Ohhashi T (1993) Effects of collagenase, elastase and hyaluronidase on the mechanical properties of isolated dog jugular veins. Am J Physiol 264 (Heart Live Physiol 33) (in press)
10. Burton AC (1954) Relation of structure to function of the tissues of the wall of blood vessels. Physiol Rev 34:619–642
11. Dobrin PB (1978) Mechanical properties of arteries. Physiol Rev 58:397–460
12. Fung YC (1981) Biomechanics: mechanical properties of living tissues. Springer-Verlag, New York, pp 261–301
13. Franklin KJ (1937) A monograph on veins. Charles C Thomas, Springfield, pp 35–75
14. Shepherd JT, Vanhoutte PM (1975) Veins and their control. Saunders, Philadelphia, pp 210–238
15. Pollack AA, Wood EH (1949) Venous pressure in the saphenous vein at the ankle in man during exercise and changes in posture. J Appl Physiol 1:649–662
16. Stehbens WE (1979) Hemodynamics and the blood vessel wall. Charles C Thomas, Springfield, pp 3–74, 94–104
17. Burch GE (1950) Primer of venous pressures. Lea and Febiger, Philadelphia
18. Ludbrook J, Beales G (1962) Femoral valves in relation to varicose veins. Lancet 1:79

Venous Capacitance Changes in Congestive Heart Failure and Exercise

JOHN V. TYBERG and SANDRA E. BAKER[1]

Abstract. A pressure–volume model of the peripheral circulation is presented with a discussion of the effects of congestive heart failure on the venous capacitance bed. The model is also used as the basis for a new interpretation of the hemodynamics of the "muscle pump".

Key words: Congestive heart failure—Venous capacitance—Exercise—Muscle pump

Introduction

The venous system remains the least well understood element in the circulation, much more being known about the heart, the arterial system, and the microcirculation. Furthermore, some of what is thought to be understood may be, if not incorrect, subject to serious misinterpretation. In the latter category, particularly, may be the concept of "venous return". There can be no doubt that venous return is a very useful concept when analyzing the beat-to-beat circulatory changes that accompany standing up or a rapid, deep inspiration. The difficulty comes when "an increase in venous return" is used to denote the steady-state condition caused by the venoconstriction-induced translocation of blood from the peripheral to the central circulations. In congestive heart failure—when cardiac output is known to be depressed—steady-state venous return *cannot* be increased [1].

Also, the analogy of the emptying lung [2] has led to an emphasis on right atrial (RA) pressure and venous resistance as the principal determinants of venous return. In this formulation, it is the reduction in RA "back pressure" which allows venous return to increase. However, this inverse relationship between RA pressure and venous return (cardiac output) has been interpreted alternatively by Levy, who suggested that it is the increase in cardiac output

---

[1] Departments of Medicine and Medical Physiology, The University of Calgary, 3330 Hospital Drive NW, Calgary, Alberta T2N 4NI, Canada

that *causes* the decrease in RA pressure (i.e., in a two-compartment system, increasing cardiac output depletes the venous reservoir) [3]. (This is equivalent to Rowell's multiple-compartment, flow-dependent redistribution of volume as cardiac output increases [4].) Although this "chicken-and-egg" problem cannot be resolved conclusively because both interpretations are internally consistent, recently we proposed a model [5] to support Levy's interpretation of the inverse relation between RA pressure and cardiac output [3] and we interpreted recent data in the context of that model.

An Alternative Model of the Peripheral Circulation

The fact that an increase in cardiac output causes a decrease in venous and RA pressures becomes clear from an examination of Fig. 1. In this figure, the idealized human circulatory values chosen by Levy [3] are used and, for simplicity, systemic vascular resistance has been assumed to be constant and

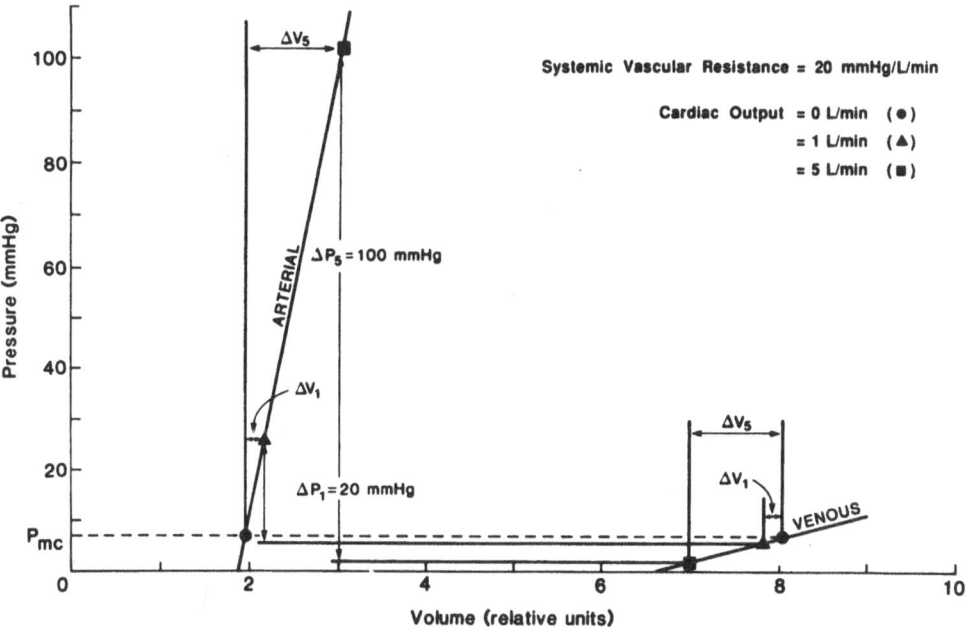

FIG. 1. Arterial and venous pressure–volume relations. For simplicity, linearity has been assumed and the slopes conform to the ratio of arterial-to-venous compliances as given by Levy [3]. ΔP_1, ΔP_5, ΔV_1, and ΔV_5 represent the respective incremental pressures and volumes associated with cardiac outputs of 1 and 5 l/min. The volume axis is relative but chosen to indicate that ≈70% of the blood is normally located in the venous circulation. Note that, since the volume of blood is conserved, the volume that is removed from the veins is the same as that added to the arteries, as cardiac output increases. From [5] with permission of Mosby-Year Book, Inc.

equal to 20 mmHg/l per min. The horizontal scale is defined only in relative units and is based on the observation that approximately 70% of the blood is contained within the venous system at normal cardiac outputs and pressures [6]. When cardiac output is zero, the pressures in the idealized venous and arterial compartments are equal and are assumed to be 7 mmHg. This pressure is defined as mean circulatory pressure (P_{mc}) [6], sometimes termed mean circulatory filling pressure or mean systemic filling pressure. Mean circulatory pressure is the value of pressure everywhere in the circulation after the heart has stopped beating (if flow is zero, there can be no pressure gradients in a resistive circuit and, therefore, all pressures must be equal). (It is not simple to determine this value experimentally [7].) Mean circulatory pressure is also a measure of "the fullness of the circulation" [7], a measure of the circulating blood volume relative to the "tone" of the vasculature. It will be shown that, as cardiac output increases, mean circulatory pressure is the value from which arterial pressure increases and venous pressure decreases.

For simplicity and convenience, Levy assumed that the ratio of venous to arterial compliance was 19:1 [3]. While this value is lower than most of those in the literature [8], the difference may be explained in part by the curvilinearity of the arterial pressure–volume relation: the real curve is nonlinear and concave upwards (i.e., pressure increases more rapidly than volume) and experimental determinations of the venous-to-arterial compliance ratio have been made using an incremental compliance (i.e., the inverse of the slope of the curve over a limited segment of the physiologic range). It is to be expected that the tangent drawn to the arterial pressure–volume curve (i.e., the slope) at a normal value of arterial pressure will be steeper than the average compliance, which is the line drawn from the level of mean circulatory pressure to physiologic arterial pressure. Since the higher, average value of arterial compliance was used, the venous-to-arterial ratio naturally became lower.

Now consider what would happen within the previously arrested and equilibrated circulation when the heart began to beat, producing a cardiac output of 1 l/min. Arterial pressure would fall, until an arterial-to-venous pressure gradient, just sufficient to drive a flow of 1 l/min back through the systemic vascular resistance into the venous compartment, had been produced. (The value of this pressure gradient would be 20 mmHg [1 l/min × 20 mmHg/l per min = 20 mmHg]). This condition would constitute a new steady state. For the pressure within a compliant arterial reservoir to increase and that within a compliant venous reservoir to decrease, the heart would have had to translocate an incremental volume of blood (ΔV_1) to the arterial reservoir from the venous reservoir (see Fig. 1). Because of the raito of the compliances which has been assumed (i.e., 1:19), this transfer of blood would raise arterial pressure by 19 mmHg (from 7 mmHg to 26 mmHg) and lower venous pressure by 1 mmHg (from 7 to 6 mmHg). Now consider the results of increasing cardiac output to 5 l/min. A new steady state would require that the arterial-to-venous pressure gradient be equal to 100 mmHg. To accomplish this, more blood would have to be translocated (ΔV_5) to the arteries from the veins. This translocation of blood would produce normal mean values of arterial and venous pressures (102 and 2 mmHg, respectively).

By definition, "capacitance" means the volume a vessel contains at a given pressure; therefore, capacitance can be changed by either a change in the *slope* of the pressure–volume relation (i.e., the reciprocal of the compliance) or by a change in the *position* of the relation [6]. Changes in venous capacitance tend to be dominated by changes in the position of the relation, that is, by changes in the so-called unstressed volume (the unstressed volume is the volume that would exist at a venous pressure of zero if the quasi-linear portion of the venous pressure-volume relation were extrapolated to the zero-pressure axis [6].) It does not *really* exist, in that all direct studies of vascular pressure–volume relations indicate that the relation is curvilinear and tends to parallel the volume axis near zero pressure [9]. Nonetheless, it has proved to be a useful parameter with which to characterize the position of the pressure–volume relation. True changes in venous compliance can be demonstrated [10], but these are small, relative to the changes in unstressed volume. They are also smaller than previous models of the vasculature would have predicted. These models were based on the assumption that the pressure–volume relations of the vasculature, like the voltage–charge relations of electrical capacitances, pass through the origin [11,12]. The small changes in compliance which have been demonstrated do not appear to be larger than those to be expected from a priori geometric considerations (i.e., if one assumes that the relative change in volume is constant for a given pressure change, then the slope of the pressure–volume relation [the reciprocal of compliance] must decrease in proportion to the increase in unstressed volume). Although the small slope changes in the directly measured intravascular pressure–diameter relation are unequivocal [10], the slopes of pressure–volume relations described by less direct techniques may be less physiologically significant in that they are model-dependent and sometimes involve a lumped capacitance in which the contribution of a given vascular element (e.g., venules) might be less at lower outlet pressures than at higher pressures. Finally, from a teleologic viewpoint, Greenway has suggested that a change in unstressed volume is preferable to a change in compliance in that the effect of a change in compliance would tend to diminish at low venous pressures and in circumstances under which translocation of volume might be most beneficial [13].

The Veins in Congestive Heart Failure

As implied at the outset, it has been widely assumed that an important component of the pathophysiology of congestive heart failure is venoconstriction. However, direct experimental evidence has been lacking until recently. Using the increasingly accepted pacing-induced canine model of chronic congestive heart failure [14,15] and classic indicator-dilution methods, Ogilvie and Zborowska-Sluis demonstrated that heart failure results in a profound leftward shift in the relation between mean circulatory filling pressure and total blood volume [16]. The decrease in unstressed volume was almost 50% and the changes in compliance were of no statistical or physiologic significance. This strongly suggests that chronic congestive heart failure involves profound constriction of the venous capacitance vasculature.

We have used the model of acute heart failure developed by Smiseth and Mjøs in which the canine left coronary artery is incrementally embolized by 50 μm microspheres until left ventricular (LV) end-diastolic pressure reaches 20 mmHg (at that time, cardiac output is usually reduced by about 50%) [17]. We studied the effects of three different vasoactive agents: nitroglycerin, enalaprilat (a parenteral angiotensin-converting enzyme inhibitor), and hydralazine. In order to distinguish effects on venous capacitance from arteriolar effects, we used a modification of Tc99m-blood pool scintigraphy [18,19] to measure changes in total vascular volume and administered each drug in equihypotensive dosages (i.e., in each case, mean arterial blood pressure was reduced by 20%). We reasoned that that this procedure would result in equal effects on the arteriolar circulation, which largely determines vascular resistance, and that changes in total vascular volume could, for the most part, be ascribed to changes in venous capacitance, since ≈70% of the blood is contained in the venous vasculature [6]. Relative to our control observations, this model of heart failure produced a 15% reduction in total vascular (primarily venous) unstressed volume. Then, in 3 different groups of dogs, nitroglycerin, enalaprilat, and hydralazine were given. Relative to the value of unstressed volume observed during heart failure, nitroglycerin increased unstressed volume by ≈30%, enalaprilat by 11%, and hydralazine by 0% [20]. These observations showed that vasodilators, administered in a dosage to produce equal arteriolar effects, can have vastly different effects on the venous capacitance bed. As widely recognized, nitroglycerin has a relatively large effect on venous capacitance, enalaprilat clearly dilates the veins but is only ≈$\frac{1}{3}$ as effective as nitroglycerin, and, consistent with textbook teaching, hydralazine has no effect on venous capacitance even though its effects on the arteriolar bed were equal. Effects on LV end-diastolic pressure were consistent with the venous effects: nitroglycerin lowered pressure the most, enalaprilat was intermediate, and there was no significant effect with hydralazine. Thus, for equal effects on the arteriolar bed, therapeutic vasodilators may have very disparate effects on venous capacitance and it is the venous effects which determine changes in cardiac filling pressure.

Speculations on the Nature of the "Muscle Pump"

The name "muscle pump" suggests that the action of contracting skeletal muscles on the peripheral circulation is to alter the relation between pressure and flow. Indeed, as demonstrated by Stegall [21] who showed that the working muscles of the lower limb can more than overcome the considerable hydrostatic gradient in an upright individual, the rhythmic contraction of skeletal muscles around deep veins with one-way valves can effectively pump blood out of the arterial tree into the venous system. However, certain other evidence also suggests that an important aspect of the circulatory effect of muscle contraction is to alter vascular pressure–volume relations (i.e., to alter vascular capacitance). Guyton et al. concluded "that skeletal muscle activity compresses the intramuscular and intra-abdominal vessels and thereby increases the mean circulatory pressure . . . this in turn translocates blood into the heart and . . . the heart then responds in conformity with Starling's principle of cardiac adaptation to increase the cardiac output [11,22]. Bevegård and Shepherd [23] clearly demonstrated a decrease in upper-limb venous capacitance during lower-limb exercise. Recently, Flamm et al. performed an elegant study of blood volume redistribution in human volunteers during a period of bicycle exercise [24]. They showed that blood was shifted, both from the exercising muscles and from the abdominal viscera, to the central compartment, in that the blood volumes of both heart and lungs increased as exercise increased in intensity.

Recent work by Sheriff et al. [25] provides the basis for a more detailed interpretation of the changes in vascular capacitance induced by skeletal muscle contraction. In dogs with chronic congestive heart failure which had undergone complete heart block and instrumentation to measure aortic flow and aortic and central venous pressures, heart rate (and thus cardiac output) was controlled by ventricular pacing. At rest, when cardiac output was changed, central venous pressure changed inversely as Guyton et al. [26] and Levy [3] had showed. When the dog began exercising by running on a horizontal treadmill, changing the cardiac output again caused the central venous pressure to change inversely. However, in this case the whole relation was shifted upward and rightward (i.e., at a given cardiac output, central venous pressure was greater), as the data of Bevegård et al. [27] in exercising patients had also shown. When the intensity of exercise was increased by increasing the angle of inclination of the treadmill, Sheriff et al. [25] demonstrated a further upward and rightward shift in the central venous pressure-cardiac output relation.

As shown in Fig. 2 [3], when cardiac output equals zero, the values of central venous pressure and arterial pressure (by definition) equal mean circulatory pressure. As cardiac output is increased, arterial pressure increases, central venous pressure decreases, and the difference between the two, the arterial–venous pressure gradient, increases. This is a simple example of the Ohm's-law behavior of the systemic circulation (i.e., the arterial–venous pressure gradient is directly proportional to the total flow through the resistance) and the relative flow-dependent changes in arterial and venous pressures are inversely proportional to their respective compliances. (It should be emphasized that these arterial and venous pressure–flow relations characterize and may be interpreted as properties of the systemic circulation, since they would be exactly the same if the heart had been completely replaced by a pump-oxygenator.) Although, in Levy's experiment, this total flow was equal to the cardiac output, which in turn was equal to the pump output, the flow through the resistance does not

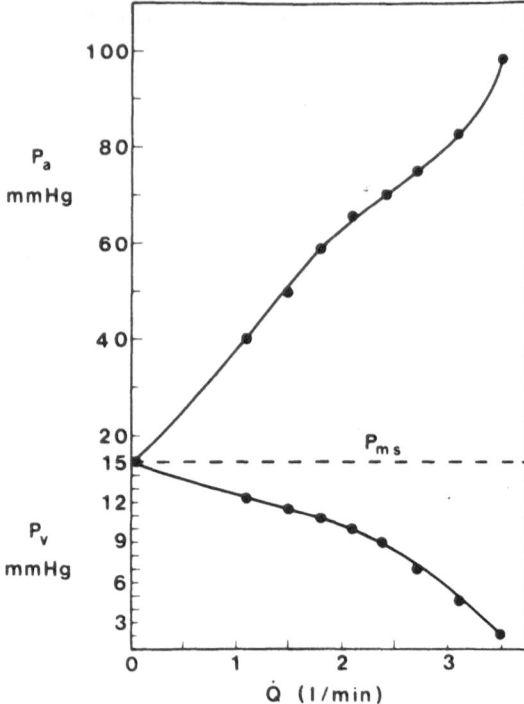

FIG. 2. Arterial (P_a) and venous (P_v) pressure–cardiac output ($\dot{Q}$) relations. As cardiac output was increased, arterial pressure increased and venous pressure decreased. (Cardiac output was controlled by a right-heart bypass.) When cardiac output was zero, both pressures were equal and, by definition, equal to mean circulatory pressure (equivalent to mean systemic pressure, P_{ms}). The difference between P_a and P_v is approximately proportional to cardiac output. (From [3] with permission of the American Heart Association.)

have to be equal to the cardiac output, as we will point out below. Sheriff et al. [25] measured arterial pressure as well as central venous pressure and cardiac output and, when the data which they recorded after autonomic blockade are plotted as Levy did (see Fig. 2), the results are very interesting (see Fig. 3). If one assumes a slight degree of curvilinearity in both curves, one can imagine that the resting arterial and venous pressure-flow relations can be extrapolated to converge at a pressure of ≈10 mmHg (i.e., P_{mc}), when cardiac output is zero. However, if one assumes that the shapes of the arterial and venous curves do not change greatly during exercise, it seems extremely unlikely that the post-exercise arterial and central venous pressure–cardiac output relations can be made to converge at a cardiac output of zero. That is, after exercise the pressure–cardiac output relations appear to be shifted rightward. Perplexingly, this seems to imply that the cardiac output values were too high. How might this have been possible?

A possible explanation is suggested by Fig. 4. Normally the heart pumps 5 l/min out of the systemic venous system into the arterial system. Given a normal systemic conductance of 0.05 l/min per mmHg, the difference between mean arterial and central venous pressures will be 100 mmHg (top panel). An example of the effects of having a pump in parallel to the heart is illustrated in the middle panel. Here it is assumed that the heart and peripheral circulation are unchanged and that a cardiac-assist device capable of delivering 2 l/min

FIG. 3. Arterial (*above*) and central venous (*below*) pressure–cardiac output relations in dogs at rest (*filled circles*) and running at 4 mph on a horizontal (0%-grade) treadmill (*open circles*). Reflexes were blocked by hexamethonium, and cardiac output was controlled by pacing after atrioventricular block. Notice that, during exercise, the curves appear to be shifted to the right and extrapolation to the pressure axis (cardiac output = 0 l/min) appears to be impossible (see Fig. 2). (Plotted from the data of Sheriff et al. [25], with permission of the American Physiological Society)

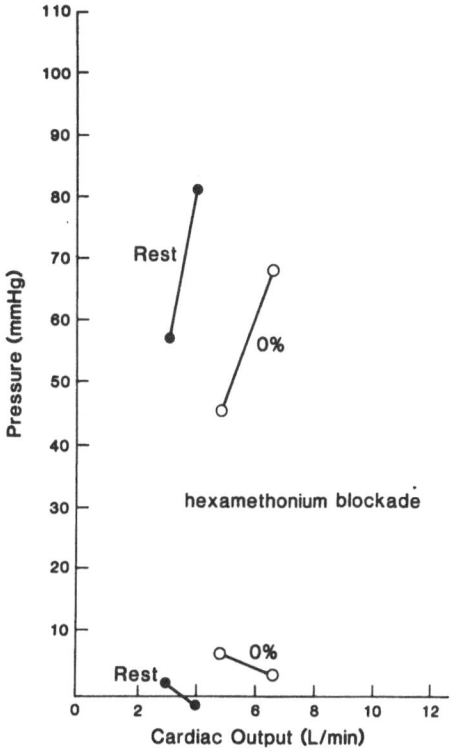

has been connected between the systemic veins and the arteries. Since the peripheral conductance has not changed and the sum of the outputs (that of the heart plus that of the cardiac-assist device) is 7 l/min, the difference between arterial and central venous pressure will be 140 mmHg. Finally, consider a different example of a pump situated in parallel with the heart (note that the pump is also in parallel with the lumped nonworking systemic vascular conductance). In this case, the device pumps blood *from the arteries to the veins*, as the muscle pump does. Since the peripheral conductance has not changed and the algebraic sum of the outputs (that of the heart plus that of the muscle pump) is 3 l/min, the difference between arterial and central venous pressures will be 60 mmHg. Similar to the results of muscular exercise, this example manifests an apparently low peripheral resistance (the arterial–venous pressure gradient divided by the cardiac output = [60 mmHg]/[5 l/min] = 12 mmHg/l per min), even though the conductance of the systemic vascular load was constant in all 3 cases and equal to a normal systemic conductance at rest. The same analysis would apply if cardiac output were 20 l/min and the muscle pump, 17 l/min (the net flow to the nonworking vasculature would also equal 3 l/min), as might be true in a well-conditioned athlete. Thus, the result of this analysis is to suggest that cardiac output may not always be a true measure of the flow

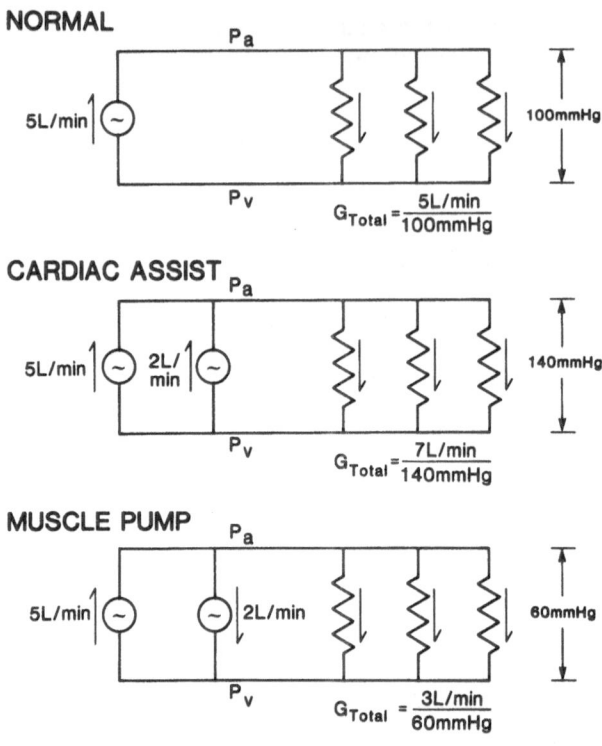

Fig. 4. Schematic diagrams of the normal circulation (*top panel*), the circulation after the addition of a cardiac-assist device (*middle panel*), and after the addition of a muscle pump (*bottom panel*). In the normal circulation (*top panel*) the heart pumps a flow of 5 l/min into a vascular system with a conductance (G) of 0.05 l/min per mmHg. This results in an arterial–central venous pressure difference ($P_a - P_v$) of 100 mmHg. The middle panel illustrates the effect of the addition of a cardiac-assist device which pumps 2 l/min. The conductance is unchanged and the sum of the outputs of the heart and cardiac-assist device equals 7 l/min. $P_a - P_v$ therefore equals 140 mmHg. In the bottom panel, a muscle pump which pumps 2 l/min from the arteries to the veins is added. The measured cardiac output is still 5 l/min but the equivalent output is 3 l/min. Since the conductance is unchanged, $P_a - P_v = 60$ mmHg. (The vascular conductance of resting skeletal muscles was neglected in the top and middle panels for simplicity)

through the system, the true flow being greater (Fig. 4, middle panel) or less than the cardiac output (bottom panel).

This model may seem inconsistent with the observations of Stegall, who demonstrated the very substantial power requirements (i.e., ≈30% of the total energy cost of the circulation) of the leg-muscle pump during running [21]. As a result of an involuntary Valsalva-like maneuver, abdominal pressure rises during running and this potential-energy barrier must be overcome in order to return blood to the right atrium (RA). If the leg-muscle pump delivers a large

FIG. 5. The data from Fig. 3 have been replotted to show the hypothetical effects of the muscle pump. We have assumed that the output of the muscle pump (i.e., the horizontal difference between the arterial rest-vs-exercise pressure–flow curves, ≈2.5 l/min) negated a fraction (2.5/5.8 × 100 = 43%) of the cardiac output (≈5.8 l/min). Note that extrapolations of the venous pressure–cardiac output curves suggest that exercise increased P_{mc} by 2 mmHg. (Data from Sheriff et al. [25]. with permission of the American Physiological Society)

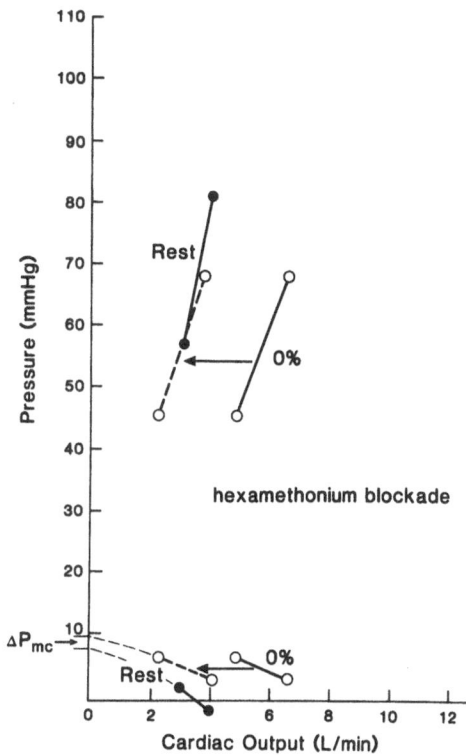

flow against a large (venous) pressure head, the energy cost of this pump would be substantial. During running, the effect of the leg-muscle pump is to return a large fraction of the total cardiac output to the RA, regardless of any intervening potential-energy barrier (i.e., elevated abdominal pressure). (This discussion suggests that the hemodynamics of repetitive upper-limb exercise might be different, in that no potential-energy barrier is interposed between the working muscle and the RA.)

The foregoing analysis suggests that, in order to express arterial and venous pressures as functions of the net venous-to-arterial flow, the exercise curves in Fig. 3 should be shifted leftward from their original positions (see Fig. 5). To estimate the magnitude of the correction from the existing data, we have used the difference in cardiac outputs (≈2.5 l/min) at an intermediate level of arterial pressure (≈60 mmHg), a region of the plot where both rest and exercise data exist and where the arterial pressure–flow relations are parallel. This analysis implies that 40% or more of the cardiac output during 0%-grade exercise (≈5.8 l/min) was "negated" by the muscle pump and that the *net flow* was only 3.2 l/min. The venous curves must also be shifted leftward and the result suggests that P_{mc} may have increased slightly (Fig. 5). Performing this analysis on Sheriff's data from unblocked animals is also interesting (Fig. 6). An analysis of these data suggests that the muscle-pump flow is also ≈2.2 l/min

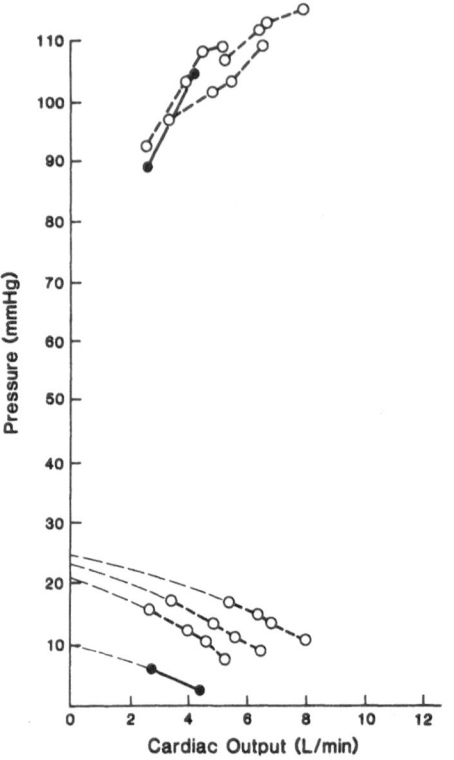

FIG. 6. Arterial (*above*) and central venous (*below*) pressure–cardiac output relations in dogs at rest and running on a treadmill at 4 miles per h at various grades (0%, 10%, and 20%). As in Fig. 5, exercise curves have been "corrected" and shifted leftward, consistent with the assumption that the muscle pump negated 40% of the cardiac output. Note that extrapolations of the venous pressure–cardiac output curves suggest that exercise increased P_{mc} by at least 10 mmHg. (Plotted from data of Sheriff et al. [25] with permission of the American Physiological Society)

and, similarly, equivalent to about 40% of the cardiac output. Conspicuously different, however, is the apparent venous effect on P_{mc}. P_{mc} appears to increase by $\approx$10 mmHg during 0%-grade exercise and to increase further when the angle of inclination of the treadmill was increased. Comparison of the blocked and unblocked states might imply that most of the changes in venous capacitance are dependent on neurohumoral control and, therefore, may be splanchnic in origin. However, this suggestion—that the "volume-displacement" component of the muscle pump is unimportant—must be reconciled with the direct measurements of muscle-blood-volume decrease ($\approx$20%) in exercising human volunteers by Flamm et al. [24].

This general analysis might be a useful approach with which to study the hemodynamics of isometric exercise as well. However, it seems obvious that the behavior of the muscle pump during isometric exercise must be quantitatively, if not also qualitatively, different. When muscles contract nonrepetitively, the "pump component" (i.e., the negation of part of the cardiac output) may be much less and the "volume-displacement component" (i.e., the increase in P_{mc}), relatively greater.

In summary, as suggested by the both upward and rightward shifts in the central venous pressure–cardiac output relation (Fig. 3), the circulatory adjustments of muscular exercise might have significant venous capacitance and "muscle pump" components. Plots of arterial and central venous pressures versus cardiac output might be used to discriminate one component from the other, if one assumes that the muscle pump translocates blood from the arteries to the veins in a parallel circulation and, thus, negates a fraction of the cardiac output.

Acknowledgments. Dr. Tyberg is a Medical Scientist of the Alberta Heritage Foundation for Medical Research (AHFMR, Edmonton). Work described in this article was supported in part by Grants-in-Aid from the Alberta Heart and Stroke Foundation (Calgary) and the Medical Research Council of Canada (Ottawa). The editorial assistance of Naomi M. Anderson, PhD, is gratefully acknowledged.

References

1. Packer M (1990) Editorial: Abnormalities of diastolic function as a potential cause of exercise intolerance in chronic heart failure. Circulation 81[Suppl III]:III-78-III-86
2. Permutt S, Caldini P (1975) Regulation of cardiac output by the circuit: venous return. In: Baan J, Noordergraaf A, Raines J (eds) Cardiovascular system dynamics. Massachusetts Institute of Technology, Boston, pp 465-479
3. Levy MN (1979) The cardiac and vascular factors that determine systemic blood flow. Circ Res 44:739-747
4. Rowell LB (1993) Human cardiovascular control. Oxford University Press, New York Oxford, p 45
5. Tyberg JV (1992) Venous modulation of ventricular preload. Am Heart J 123: 1098-1104
6. Rothe CF (1983) Reflex controls of veins and vascular capacitance. Physiol Rev 63:1281-1341
7. Rothe CF (1993) Mean circulatory pressure: Its meaning and measurement. J Appl Physiol 74:499-509
8. Rothe CF (1983) Venous system: Physiology of the capacitance vessels. In: Shepherd JT, Abboud FM (eds) The cardiovascular system. Volume III. Peripheral circulation and organ blood flow, Part 1. American Physiological Society, Bethesda, pp 412-413
9. Katz AI, Chen Y, Moreno AH (1969) Flow through a collapsible tube: Experimental analysis and mathematical model. Biophys J 9:1261-1279
10. Shoukas AA, Bohlen HG (1990) Rat venular pressure-diameter relationships are regulated by sympathetic activity. Am J Physiol 259:H674-H680
11. Guyton AC, Coleman TG, Granger HJ (1972) Circulation: Overall regulation. Ann Rev Physiol 34:13-46
12. Greenway CV (1981) Simple model of the circulation. Physiologist 24:63-67
13. Greenway CV, Seaman KL, Innes IR (1985) Norepinephrine on venous compliance and unstressed volume in cat liver. Am J Physiol 284:H468-H476
14. Whipple GH, Shefield LT, Woodman EG, Theophilis C, Friedman S (1962) Reversible congestive heart failure due to rapid stimulation of the normal heart. Proc N Engl Cardiovasc Soc 20:39

15. Coleman HN, Taylor RR, Pool PE, Whipple GH, Covell JW, Ross J Jr, Braunwald E (1971) Congestive heart failure following chronic tachycardia. Am Heart J 81: 790–798

16. Ogilvie RI, Zborowska-Sluis D (1992) Effect of chronic rapid ventricular pacing on total vascular capacitance. Circulation 85:1524–1530

17. Smiseth OA, Mjøs OD (1982) A reproducible and stable model of acute ischaemic left ventricular failure in dogs. Clin Physiol 2:225–239

18. Robinson VJB, Smiseth OA, Scott-Douglas NW, Smith ER, Tyberg JV, Manyari DE (1990) Assessment of the splanchnic vascular capacity and capacitance: A new application of equilibrium blood-pool scintigraphy. J Nucl Med 131:154–159

19. Smiseth OA, Manyari DE, Lima JA, Scott-Douglas NW, Kingma I, Smith ER, Tyberg JV (1987) Modulation of vascular capacitance by angiotensin and nitroprusside: A mechanism of changes in pericardial pressure. Circulation 76:875–883

20. Wang Y, Scott-Douglas NW, Manyari DE, Tyberg JV (1991) Different effects of vasoactive agents on vascular capacitance in dogs with experimental heart failure (abstract). FASEB J 5:A776

21. Stegall HF (1966) Muscle pumping in the dependent leg. Circ Res 19:180–190

22. Guyton AC, Douglas BH, Langston JB, Richardson TQ (1962) Instantaneous increase in mean circulatory pressure and cardiac output at onset of muscular activity. Circ Res 11:431–441

23. Bevegård BS, Shepherd JT (1965) Changes in tone of limb veins during supine exercise. J Appl Physiol 20:1–8

24. Flamm SDT, Moore RL, Keech FM, Maltais F, Ahmad M, Callahan R, Dragotakes S, Alpert N, Strauss HW (1990) Redistribution of regional and organ blood volume and effect on cardiac function in relation to upright exercise intensity in healthy human subjects. Circulation 81:1550–1559

25. Sheriff DD, Zhou XP, Scher AM, Rowell LB (1993) Dependence of cardiac filling pressure on cardiac output during rest and dynamic exercise in dogs. Amer J Physiol (in press)

26. Guyton AC, Lindsey AW, Abernathy B, Richardson T (1957) Venous return at various right atrial pressures and the normal venous return curve. Am J Physiol 189:609–615

27. Bevegård BS, Jonsson B, Karlof I, Lagergren H, Sowton E (1967) Effect of changes in ventricular rate on cardiac output and central venous pressures at rest and during exercise in patients with artificial pacemakers. Cardiovasc Res 1:21–33

Effects of Vasodilators on Venous Distensibility in Humans

Tsutomu Imaizumi, Shin-ichi Ando, and Akira Takeshita[1]

Abstract. We examined the effects of sodium nitroprusside (SNP), atrial natriuretic peptide (ANP), and nitroglycerin on venous distensibility (VD) in human forearms. In healthy volunteers ($n = 8$), using a water plethysmograph, we measured VD from changes in forearm volume when transmural venous pressure (TMVP) was increased stepwise by 5–10 mmHg increments up to 30 mmHg during intra-arterial infusion of saline, ANP, and SNP. The doses of ANP and SNP were chosen to double the baseline forearm blood flow obtained during saline infusion. ANP and SNP increased VD to the same extent compared with that during saline infusion ($P < 0.05$). Sublingual nitroglycerin similarly increased VD in healthy humans ($P < 0.05$). Our results suggest that ANP, SNP, and nitroglycerin increase venous distensibility in forearms of healthy humans to a similar extent.

Keywords: Atrial natriuretic peptide—Nitroglycerin—Water plethysmograph—Sodium nitroprusside

Introduction

Many vasodilators are used in clinical medicine. Among them, sulingual nitroglycerin (NG) and intravenous sodium nitroprusside (SNP) are the most frequently used vasodilators. These drugs dilate the vein as well.

Recently, atrial natriuretic peptide (ANP) was discovered, and it will be introduced into clinical medicine as a new vasodilator [1]. However, little is known about the effects of ANP on the vein in humans. Therefore, in this study we aimed to determine the effects of NG, SNP, and ANP on venous distensibility in humans, and to compare the effects.

---

[1] Research Institute of Angiocardiology and Cardiovascular Clinic, Faculty of Medicine, Kyushu University, 3-1-1 Maidashi, Higashi-ku, Fukuoka, 812 Japan

Subjects and Methods

Subjects

The effects of ANP and SNP were studied in eight healthy male volunteers, whose ages ranged from 18 to 23 years (20.0 ± 0.6 years, mean ± SEM). All subjects were normotensive (mean blood pressure, 95.2 ± 2.9 mmHg) and on no cardiovascular drug. The study of the effect of NG was also conducted in healthy subjects. The study protocol was thoroughly explained and informed consent was obtained from each subject.

General Procedures

The study was done in the supine position. Under local anesthesia with 2% procaine, the left brachial artery was cannulated with a 20 gauge intravascular over-the-needle teflon cannula (Quick-Cath, Travenol Laboratories, Baxter Healthcare Corporation, Deerfield, Illinois, USA), which was used for drug infusion as well as for arterial pressure recording. Arterial pressure was recorded by connecting the arterial line to a pressure transducer (Viggo-Spectramed, Oxnard, California, USA) using a three-way stopcock. The arterial line was kept open by infusing heparinized saline (0.2 ml/min) while no drug was being infused. Arterial cannulation was done for the study of ANP and SNP, but not for the study of NG. A subcutaneous vein in the middle portion of the left forearm was cannulated with the same cannula as used for the artery, which was connected to a transducer (TM1, Toyo Boldwin, Tokyo, Japan) for measurement of venous pressure. The venous line was filled with heparinized saline. The venous cannula was positioned so as to be completely enclosed in the plethysmograph box. Heart rate was obtained by counting the pulse rate for a few minutes on arterial pressure recordings.

Measurements of Changes in Forearm Volume

Changes in forearm volume were measured by a water-filled plethysmograph as described previously [2]. In brief, the left forearm was enclosed in an acrylic plastic water-filled box with a capacity of approximately 8500 ml (Asahi Seisakusho, Fukuoka, Japan) and the box filled with warm water (34°C) to a level 26 cm above the upper aspect of the forearm. A rubber membrane separated the forearm from water. Water temperature was continuously monitored by a thermometer; the change in temperature during the study was within 2°C. Changes in the height of the water surface induced by changes in forearm volume were measured by a displacement transducer (DT 4812, San-ei, Tokyo, Japan). Under these conditions, the external water pressure initially collapsed the veins, but the arterial inflow caused the venous pressure to reach a level slightly higher than that of the external water pressure. The difference between the pressure within the veins and the external water pressure

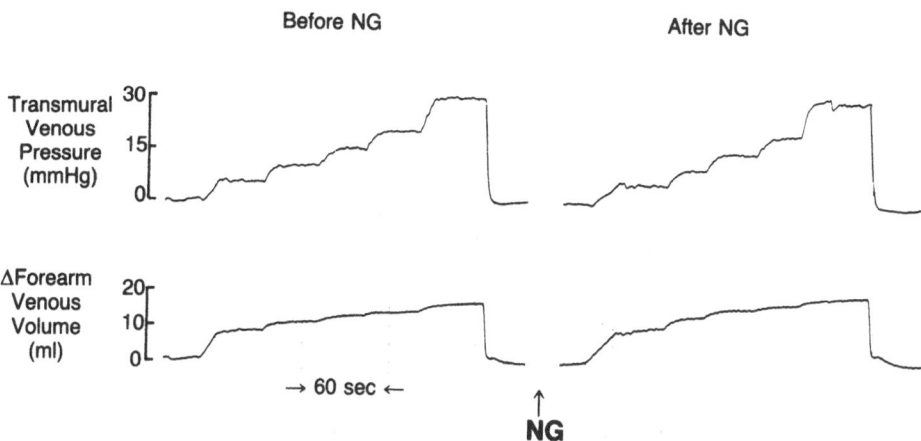

FIG. 1. Original recordings of changes in forearm venous volume (*lower trace*) and transmural venous pressure (*upper trace*). Transmural venous pressure was increased in a stepwise fashion by 5–10 mmHg increments up to 30 mmHg. Nitroglycerin (*NG*) increased changes in forearm venous volume at comparable transmural venous pressure

surrounding them is the distending or transmural venous pressure. Transmural venous pressure was measured by placing the reference level of the pressure transducer at the surface level of the water in the plethysmograph. It has been suggested that venous volume and transmural venous pressure at resting conditions are minimal and reproducible in a given subject and are similar between subjects [3].

To assess venous distensibility, we measured changes in forearm volume during stepwise increases in transmural venous pressure by 5–10 mmHg increments up to 30 mmHg by inflating a cuff on the upper arm. Transmural venous pressure was held constant for approximately 45 seconds at each step, by which time changes in forearm volume associated with increases in transmural pressure were stabilized (Fig. 1). This procedure was carried out to minimize the effect of nonuniform filling of the veins on the increase in forearm volume [4]. Changes in forearm volume under these conditions reflected changes in venous volume in the forearm enclosed in a water plethysmograph. Venous pressure-volume curves were constructed by plotting changes in forearm volume (ml/100 ml of forearm) against corresponding levels of transmural venous pressure.

Measurements of Forearm Blood Flow

Forearm blood flow was calculated from changes in forearm volume during intermittent inflation of a cuff on the upper arm. The cuff on the upper arm was inflated for 7 s and deflated for 8 s. The pressure in the venous occlusion or congesting cuff was 40 mmHg. Circulation to the hand was arrested during the determination of forearm blood flow by inflating a cuff around the wrist to suprasystolic pressure. Forearm blood flow was taken as the average of 4–6 flow measurements made at 15-s intervals.

Drug Preparation

SNP (Wako Jun-yaku, Tokyo, Japan) was dissolved in saline (2 μg/ml) immediately prior to use and stored in a black bag to prevent exposure to light. Twenty-eight amino acid alpha-human ANP (Suntory Co., Tokyo, Japan) was diluted in saline and frozen at a concentration of 5 μg/ml.

Protocol

Saline was intra-arterially infused at a rate of 0.2 ml/min. During infusion, transmural venous pressure was increased stepwise by 10 mmHg increments to 30 mmHg and changes in forearm volume were measured. Measurements of venous distensibility took approximately 3 min to complete. Then, ANP or SNP was intra-arterially infused and the same measurements were repeated. We adjusted the infusion rate of SNP (516 ± 68 ng/min) and ANP (771 ± 107 ng/min) to double the forearm blood flow at control during saline infusion. Forearm blood flow was 3.4 ± 0.5, 7.9 ± 0.9, and 8.0 ± 1.1 ml/min per 100 ml of forearm during intra-arterial infusion of saline, ANP and SNP, respectively. Forearm blood flow during ANP infusion did not significantly differ from that during SNP infusion. Infusion rates of saline, ANP, and SNP were 0.20, 0.15 ± 0.02, and 0.26 ± 0.03 ml/min, respectively. We confirmed that the difference in infusion rates had negligible effects on forearm blood flow. After completion of the study with the first drug, we waited for 10 to 30 min. After the forearm blood flow had fully returned to the baseline level, we infused the second drug and repeated the measurements. In 6 subjects, the drugs were given in the order of SNP then ANP, and in 2 subjects in the order of ANP then SNP. ANP was infused at the first drug in only 2 subjects because it took 30 min for forearm blood flow to return to baseline values after termination of the infusion. However, the results obtained in the 2 subjects in whom ANP was infused first were similar to those obtained in the other six subjects.

In order to determine the effects of NG on venous distensibility, one tablet of NG was administered sublingually. Measurements of venous distensibility were done before and 5 min after NG. After completion of the studies with the drugs, the volume of water in the acrylic box was measured. The difference in water volume with and without the forearm in the box reflected the baseline forearm volume contained in the plethysmograph.

Statistical Analysis

Values are expressed as means ± SEM in this report. Comparison of forearm blood flow was done by one-way analysis of variance (ANOVA) on repeated measurements. Venous pressure-volume curves were compared by two-way ANOVA on repeated measurements. A *P* value less than 0.05 was considered statistically significant.

Results

Intra-arterial infusion of ANP and SNP did not alter arterial pressure or heart rate. Mean arterial pressures during intra-arterial infusion of saline, ANP, and SNP were 95 ± 3 mmHg, 93 ± 3 mmHg, and 94 ± 3 mmHg, respectively. Heart rates during intra-arterial infusion of saline, ANP, and SNP were 71 ± 5 min^{-1}, 70 ± 5 min^{-1} and 69 ± 5 min^{-1}, respectively.

Both intra-arterial ANP and SNP significantly shifted the venous pressure-volume curves toward the volume axis compared to the curve during saline infusion ($P < 0.05$, ANP or SNP versus saline) (Fig. 2). The shifts of the curves caused by ANP and SNP were comparable (NS).

Nitroglycerin similarly shifted the venous pressure-volume curve toward the volume axis in healthy subjects ($P < 0.05$) (Fig. 3).

The baseline venous pressure during intra-arterial saline, ANP, and SNP infusion was 6.5 ± 0.4 mmHg, 6.5 ± 0.7 mmHg, and 6.0 ± 0.5 mmHg, respectively (NS).

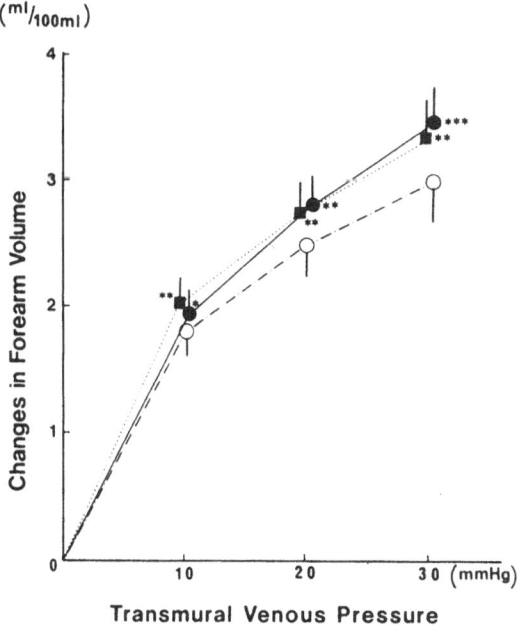

FIG. 2. Changes in forearm venous volume against changes in transmural venous pressure during intra-arterial infusion of saline (○), atrial natriuretic peptide (■), and sodium nitroprusside (●) ($n = 5$). The curve during ANP or SNP infusion was shifted towards the volume axis compared with that during saline infusion ($P < 0.05$, ANP or SNP versus saline). Asterisks indicate a significant difference from the value during saline infusion at each level of transmural venous pressure (*, $P < 0.05$; **, $P < 0.01$; ***, $P < 0.001$). Values are expressed as means ± SEM. (From [18])

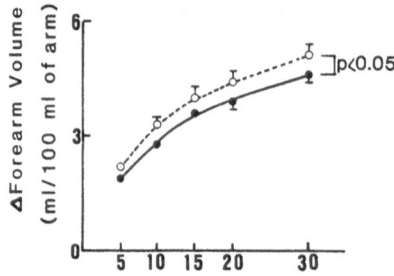

FIG. 3. Changes in forearm venous volume against changes in transmural venous pressure before (●) and after nitroglycerin (○). The curve after nitroglycerin was shifted towards the volume axis compared with that before nitroglycerin ($P < 0.05$). Values are expressed as means ± SEM

Discussion

We determined the increases in forearm volume occurring when transmural venous pressure was raised in a stepwise fashion. When transmural venous pressure was raised by increasing the cuff pressure on the upper arm, and maintained for about 45 s, forearm volume rapidly increased initially, and then leveled off (Fig. 1). This initial increase in forearm volume reflected the increase in venous capacity caused by the increase in transmural venous pressure, since the forearm volume immediately returned to the baseline level after deflation of the occluding cuff on the upper arm [2].

It has been well documented that nitroglycerin and sodium nitroprusside increase venous distensibility in humans as well as animals. In this study, we also demonstrated that intra-arterially infused SNP and sublingually administered nitroglycerin increased venous distensibility in healthy humans. Venous distensibility was examined by measuring the increases in venous capacity associated with the stepwise increases in transmural venous pressure.

Although the venodilating effects of ANP have been shown in in vitro studies [5–9], the results from in vivo studies with intravenous administration of ANP are inconsistent. It has been suggested that intravenous ANP in animals may cause venodilatation, venoconstriction, or have no effect on veins [7,10–12]. Previous studies in humans have also demonstrated no or little venodilatation evoked by intravenous ANP [13,14]. However, intravenous ANP decreases plasma volume and causes hypotension. The decrease in plasma volume and secondary neurohumoral activation would modulate the direct effect of ANP on veins [15,16]. In a previous study in humans, ANP was infused directly into a superficial vein in a hand, which caused minimal venodilatation [17]. However, it has been suggested that the response to ANP might differ among the segments of veins [5,7,8].

Our results indicate that intra-arterial ANP increased venous distensibility in human forearms (Fig. 2). The results in this study were different to those obtained by Groban et al., who reported that intravenous ANP did not alter venous distensibility [13]. Thus, our study was contradictory to their results, but it was more carefully conducted than theirs in at least one respect: they infused ANP intravenously, which caused a decrease in the central venous pressure by 4 mmHg. Such a large decrease in the central venous pressure would have caused reflex neurohumoral activation and thus may have caused venoconstriction. However, our concern was whether ANP changes venous distensibility as a result of the local effect, without neurohumoral changes. In this study we infused ANP intra-arterially, though the doses were non-physiological, and did not cause systemic hemodynamic changes and thus neuro-humoral excitation. Of course, we cannot deny the possibility that SNP and ANP might dilate different vascular segments even though the forearm vascular resistance was same during the infusion of the drugs, which would complicate the interpretation of the results.

The significance of our findings is the demonstration that intra-arterial ANP can increase forearm venous distensibility in normal men, and that the veno-dilating effect of ANP in human forearms is similar to that of SNP at the doses of the drugs that produce comparable increases in forearm blood flow.

Since the measurements of venous distensibility during infusion of ANP were done after measurements during infusion of saline, increased venous disten-sibility by ANP might have been due to stress relaxation of the veins. However, this is unlikely, since, in our previous study, we measured venous distensibility twice sequentially and the results of the first and second measurements were similar [2].

In summary, sublingual nitroglycerin, intra-arterially infused sodium nitro-prusside, and atrial natriuretic peptide increased venous distensibility in humans to a similar extent.

Acknowledgment. This study was supported by Grant-in-Aid for General Scientific Research and by a Grant-in-Aid for Scientific Research on Priority Areas from the Japanese Ministry of Education, Science and Culture.

References

1. Lai C-P, Egashira K, Tashiro H, Narabayashi H, Koyanagi S, Imaizumi T, Takeshita A (1993) Beneficial effects of atrial natriuretic peptide on exercise-induced myocardial ischemia in patients with stable effort angina pectoris. Circulation 87:144–151
2. Ito N, Takeshita A, Higuchi S, Nakamura M (1986) Venous abnormality in normotensive young men with a family history of hypertension. Hypertension 8:142–146
3. Litter J, Wood JE (1954) The volume and distribution of blood in the human leg measured in vivo: 1. The effects of graded external pressure. J Clin Invest 33:798–806
4. Brown E, Greenfield DM, Goei JS, Plassaras G (1966) Filling and emptying of the low-pressure blood vessels of the human forearm. J Appl Physiol 21:573–582

5. Faison EP, Siegl PKS, Morgan G, Winquist RJ (1985) Regional vasorelaxant selectivity of atrial natriuretic factor in isolated rabbit vessels. Life Sci 37:1073–1079

6. Labat C, Norel X, Benveniste J, Brink C (1988) Vasorelaxant effects of atrial peptide II on isolated human pulmonary muscle preparation. Eur J Pharmacol 150:397–400

7. Faber JE, Gettes DR, Gianturco DP (1988) Microvascular effects of atrial natriuretic factor: Interaction with a_1- and a_2-adrenoceptors. Circ Res 1988: 63:415–428

8. Kawai Y, Ohashi T (1989) Heterogeneity in responses of isolated monkey arteries and veins to atrial natriuretic peptide. Can J Physiol Pharmacol 67:326–330

9. Hughes AD, Nielsen H, Thom S, Martin GN, Sever PS (1987) The effect of atrial natriuretic peptide on human blood vessels. J Hypertension 5(Suppl 5):S51–S53

10. Trippodo NC, Cole FE, Frohlich ED, Macphee AA (1986) Atrial natriuretic peptide decreases circulatory capacitance in areflexic rats. Circ Res 59:291–296

11. Eliades D, Swindall B, Johnston J, Pamnani M, Haddy FJ (1989) Effects of ANP on venous pressures and microvascular protein permeability in dog forelimb. Am J Physiol 257 (Heart Circ Physiol 26):H272–H279

12. Holtz J, Stewart DJ, Elsner D, Bassenge E (1986) In vivo atrial peptide-venodilation: Minimal potency relative to nitroglycerin in dogs. Life Sci 39:2177–2184

13. Groban L, Cowley AW, Ebert TJ (1990) Atrial natriuretic peptide augments forearm capillary filtration in humans. Am J Physiol 259:H258–H263

14. Hughes A, Thom S, Goldberg P, Martin G, Sever P (1988) Direct effect of α-human atrial natriuretic peptide on human vasculature in vivo and in vitro. Clin Sci 74:207–211

15. Lappe RW, Smits JFM, Todt JA, Debets JJM, Wendt RL (1985) Failure of atriopeptin II to cause arterial vasodilation in the conscious rat. Circ Res 56:606–612

16. Pegram BL, Kardon MB, Trippodo NC, Cole FE, MacPhee AA (1985) Atrial extract: Hemodynamics in Wistar-Kyoto and spontaneously hypertensive rats. Am J Physiol 249 (Heart Circ Physiol 18):H265–H271

17. Ford GA, Eichler HG, Hoffman BB, Blaschke TF (1988) Venous responsiveness to atrial natriuretic factor in man. Br J Clin Pharmacol 26:797–799

18. Ando S, Imaizumi T, Harada S, Hirooka Y, Takeshita A (1992) Atrial natriuretic peptide increases human capillary filtration and venous distensibility. J Hypertension 10:451–457

Forearm Stiffness in Patients with Congestive Heart Failure

Masahiko Iizuka[1], Hiroshi Sato[2], Hiroshi Ikenouchi[3], Shin-ichi Momomura[3], and Takashi Serizawa[3]

Abstract. Forearm venous distensibility was studied in 24 patients with congestive heart failure using strain gauge arch plethysmography. The venous distensibility was assessed as a venous stiffness constant by fitting a forearm venous pressure-volume curve to an exponential function. Venous pressure-volume curves were clearly separated among New York Heart Association (NYHA) classes with leftward shift and increased slopes correlating with increased NYHA grade. The venous stiffness constant was elevated significantly with increased severity of symptoms (NYHA classes: $P < 0.02$) and hemodynamic parameters (pulmonary vascular resistance: $r = 0.73$, $P < 0.02$; pulmonary capillary wedge pressure: $r = 0.54$, $P < 0.02$; cardiac index: $r = 0.45$, $P < 0.03$; systemic vascular resistance: $r = 0.45$, $P < 0.03$). While the expected effect of nitroglycerin on the venous stiffness constant was demonstrated in this study, the effect of alpha human atrial natriuretic polypeptide proved minimal.

Key words: Plethysmography—Venous pressure-volume curve—Venous stiffness constant—Nitroglycerin—Alpha human atrial natriuretic polypeptide

Introduction

It is well known that venous constriction plays an important role in forming signs and symptoms of congestive heart failure by increasing preload [1,2]. The remarkable effectiveness of venodilatation therapy provides evidence supporting this assertion [3]. However, detailed and quantitative relationships between venoconstriction, hemodynamics, and clinical profiles of patients have not been

[1] First Department of Internal Medicine, Dokkyo University School of Medicine, Mibu, Shimotsuga-gun, Tochigi, 321-02 Japan
[2] Cardiovascular Institute, Tokyo, Japan
[3] Second Department of Internal Medicine, Faculty of Medicine, University of Tokyo, Tokyo, Japan

elucidated. Therefore, we studied the characteristics of forearm circulation by strain gauge arch plethysmography, and quantified venous stiffness in patients with congestive heart failure and compared it with clinical severity and hemodynamic and hormonal data [4,5].

Methods

Subjects

Twenty-four patients with congestive heart failure (mean age: 55 ± 12 years) were divided into 4 groups according to the New York Heart Association (NYHA) functional classification (I: $n = 7$, II: $n = 7$, III: $n = 5$, IV: $n = 5$). The underlying heart diseases were ischemic heart disease, dilated cardiomyopathy, and valvular heart disease. Details are given in Table 1 of [5]. All vasodilators were discontinued at least 12 hours prior to the onset of the study.

Hemodynamic and Hormonal Measurements

Hemodynamic parameters were measured by inserting a balloon thermodilution catheter, with patients lying in a supine position. In 16 patients, blood samples were taken to evaluate plasma concentrations of catecholamines and angiotensin II.

Plethysmographic Technique

Forearm plethysmography was carried out with the subject relaxed in a supine position with the right or left arm supported at a level above the heart to empty the vein, in a room where the temperature was maintained between 23° and 26°C. A venous occlusion cuff (8.5 × 33 cm) was placed around the arm just above the elbow. Another cuff was placed around the wrist. A silicone strain gauge arch (SG-24, Medasonics Inc.), 24 cm in length, was positioned so that it encircled the forearm at its widest part. The correlation between the increase in length of the strain gauge arch and that of the electrical resistance was linear within the extent of changes observed in this study. The wrist cuff was inflated to exclude circulation to the hand. Volume changes were calculated by a computer (SPG 16, Medasonics, Inc.) from those of the forearm girth on the assumption that the shape of the cross section of the forearm was constant. With the recording paper running, the cuff was rapidly inflated to 40 mmHg until the venous pressure was elevated to over 24 mmHg or until a plateau was reached. The cuff was then allowed to return to the baseline state. A series of four measurements were used to determine the data.

Calculation of Venous Stiffness Constant

The venous stiffness constant (k) was calculated from the forearm pressure-volume curve by fitting it to an exponential equation using 12 points (2–24 mmHg) ($r = 0.98 \pm 0.01$):

$$VP = a \cdot e(kv)$$

where VP is venous pressure, v is the change in forearm volume, and a is a constant. Actually, our method measures only changes in venous pressure (ΔP) and forearm volume (ΔV). Thus, absolute venous pressure and forearm volume are expressed as follows: $VP = P_0 + \Delta P$, $V = V_0 + \Delta V$, where P_0 is the initial venous pressure and V_0 is the initial absolute forearm volume.

$$P_0 + \Delta P = a \cdot e(k \cdot (V_0 + \Delta V))$$

P_0 was set to zero by arm elevation in this study.

$$P = a \cdot e(k \cdot V_0) \cdot e(k \cdot \Delta V) = c \cdot e(k \cdot \Delta V)$$

where c is a constant.

From these relationships, k, the venous stiffness constant, proved to be independent of the initial absolute forearm volume. The amount of venous volume change at a venous pressure of 20 mmHg ("V20") was also measured as the venous compliance at a single pressure point.

Effects of Drugs

The effect of nitroglycerin was studied in 8 patients with congestive heart failure (mean NYHA class: 2.57). Nitroglycerin ointment (5 cm containing about 30 mg of nitroglycerin, Vasolator, Sanwa Chemical Lab. Inc.) was applied to the chest over an area of about 5 × 5 cm. In the baseline state and after 15 and 30 min, data were obtained.

Seven patients with congestive heart failure (mean NYHA class: 2.86) were studied to determine the effect of alpha human atrial natriuretic polypeptide (hANP) which was infused continuously at a constant dose (0.1 µg/kg per min) for 30 min by an infusion pump. Forearm plethysmography was performed at 15 min intervals throughout the infusion and postinfusion periods.

Details of patients studied are given in Table 1 of [4].

Statistical Analysis

All values are presented as means ± SD. Differences among all NYHA groups were analyzed by one-way analysis of variance followed by unpaired modified Student's t-tests. Relations between k and other parameters were evaluated by linear regression analysis. P values less than 0.05 were considered to be significant.

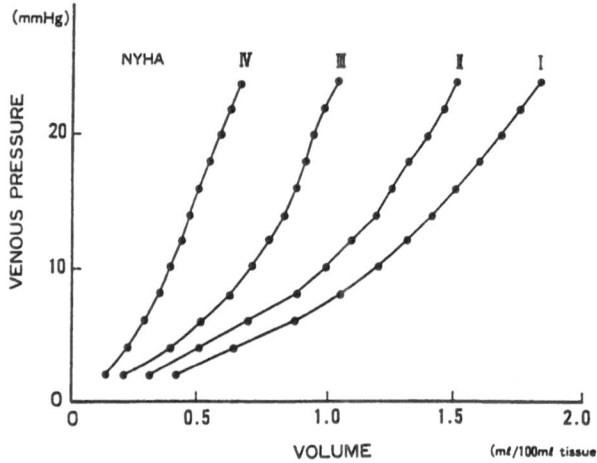

FIG. 1. Mean forearm venous pressure-volume curves of NYHA functional classes. (From [5]). *NYHA*, New York Heart Association (classes I–IV)

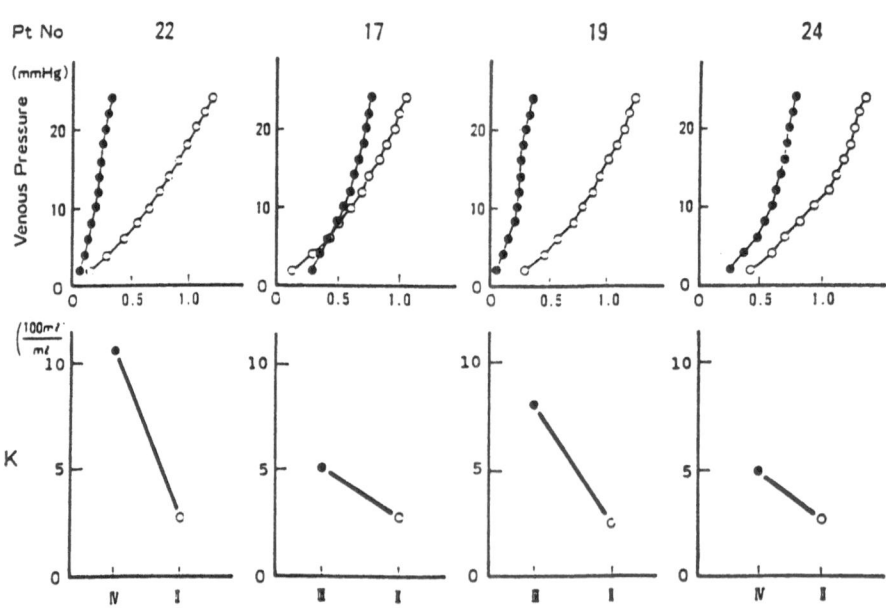

FIG. 2. Improvement of forearm venous distensibility brought about by the treatment in patients with severe congestive heart failure. ●, pretreatment data; ○, post-treatment data. *Upper panels*, pressure-volume curves; *Lower panels*, venous stiffness constants (*K*). (From [5]). *Pt No*, patient number

TABLE 1. Comparison of forearm hemodynamic parameters among NYHA functional classes.

	NYHA class				
	I	II	III	VI	
Forearm vascular resistance (mmHg·100 ml·min/ml)	80 ± 38	84 ± 56	55 ± 23	129 ± 57	NS
Forearm venous stiffness constant (100 ml/ml)	0.78 ± 0.24	1.00 ± 0.48	1.71 ± 1.12	2.32 ± 1.19	$P < 0.02$

NYHA, New York Heart Association

Results

Forearm Venous Pressure-Volume Curve and NYHA Classification

Forearm venous pressure-volume curves in each group are presented in Fig. 1 [5]. The curves are shifted to the left on the volume axis and become steeper as the NYHA grade increases. The forearm venous stiffness constant (k) and forearm vascular resistance in each group are shown in Table 1. In 4 patients, studies were repeated after clinical improvement with treatment. Changes in the venous pressure-volume curves and stiffness constant are demonstrated in Fig. 2 [5].

Venous Stiffness Constant and Hemodynamic Parameters

There was a significant negative correlation between the venous stiffness constant k and the venous compliance V20 ($r = -0.73$, $P < 0.001$). The venous stiffness constant had significant correlation to pulmonary vascular resistance ($r = 0.73$, $P < 0.001$) (Fig. 3). Less significant correlations were found with systemic vascular resistance ($r = 0.45$, $P < 0.03$), pulmonary capillary wedge pressure ($r = 0.54$, $P < 0.02$), right atrial pressure ($r = 0.51$, $P < 0.02$), and cardiac index ($r = 0.45$, $P < 0.03$). In contrast, V20 showed no significant correlation with any parameters. Detailed data are shown in Table IV of [5].

Hormonal analysis revealed a significant relationship between k and plasma norepinephrine concentration ($r = 0.64$, $P < 0.008$). V20 had a less marked relationship. Plasma epinephrine concentration had a significant correlation with k ($r = 0.56$, $P < 0.03$) but not with V20.

Effects of Drugs

Thirty minutes after nitroglycerin ointment was applied, hemodynamic changes were maximal. Forearm blood flow was slightly increased (1.7 ± 0.3 to 2.2 ±

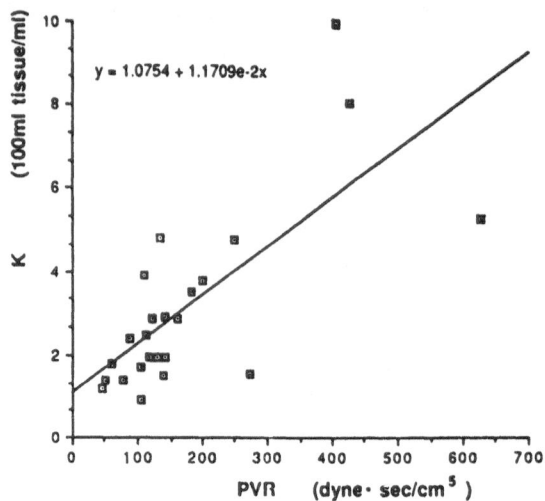

FIG. 3. Relationship between pulmonary vascular resistance (*PVR*) and forearm venous stiffness constant (*K*). (From [5])

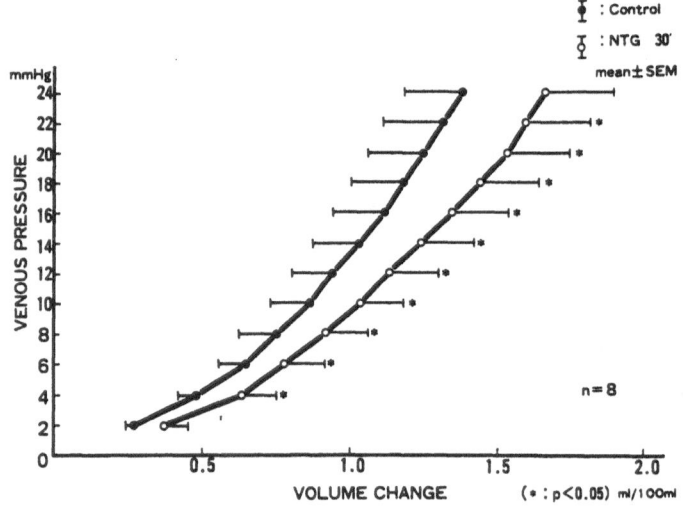

FIG. 4. Effect of nitroglycerin (*NTG*) administration on forearm venous pressure-volume curve. ●, control state; ○, state after NTG application. (From [4])

0.6 ml/100 g per min) and vascular resistance was slightly decreased (73 ± 22 to 61 ± 16 mmHg × 100 ml × min/ml); neither change was statistically significant. The forearm pressure-volume curve was shifted to the right on the volume axis (Fig. 4) [4]. The venous stiffness constant was significantly decreased (1.14 ± 0.2 to 0.88 ± 0.2, *P* < 0.04) (Fig. 5) [4].

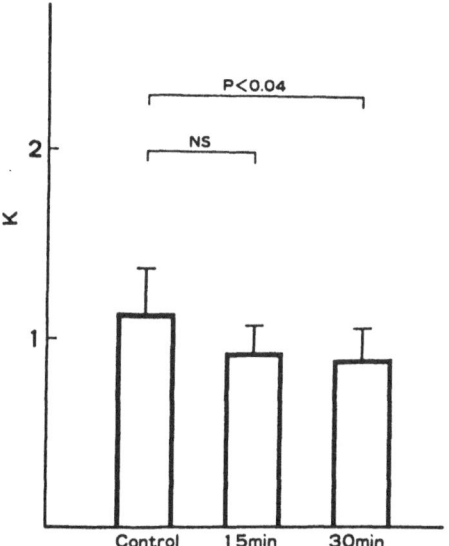

FIG. 5. Changes of the forearm venous stiffness constant (*K*) 15 and 30 min after nitroglycerin administration. (From [4])

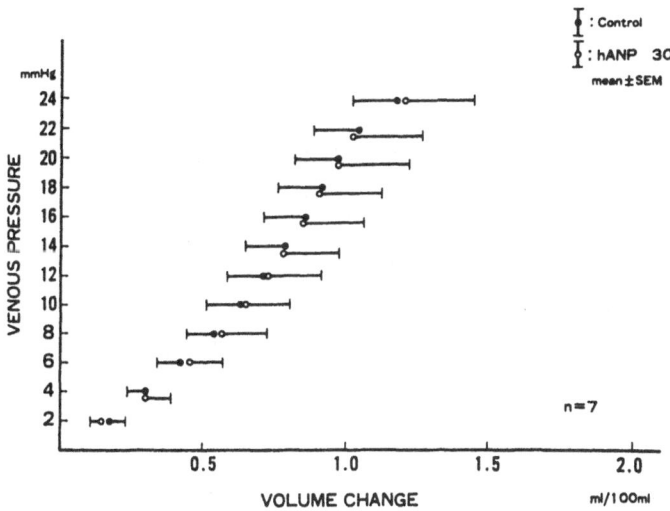

FIG. 6. Effects of alpha human atrial natriuretic polypeptide (*hANP*) on the forearm venous pressure-volume curve. ●, control state; ○, state after hANP administration. (From [4])

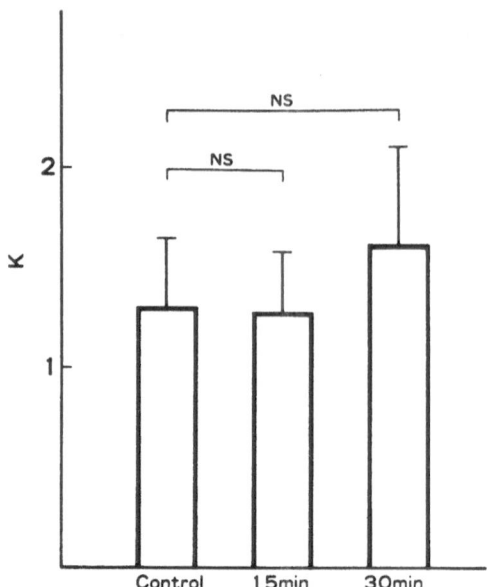

FIG. 7. Changes in the forearm venous stiffness constant (K) 15 and 30 min after administration of alpha human natriuretic polypeptide. (From [4])

Significant hemodynamic effects started 15 min after the onset of alpha human atrial natriuretic polypetide (hANP) infusion, and were sustained thereafter. Pulmonary capillary wedge pressure decreased significantly (16 ± 4 to 13 ± 4 mmHg, $P < 0.05$), right atrial pressure decreased slightly (NS), the cardiac index increased (2.7 ± 0.3 to 3.0 ± 0.3 l/min per m^2, $P < 0.05$), and systemic vascular resistance decreased significantly (1984 ± 278 to 1618 ± 212 mmHg · 100 ml · min/ml, $P < 0.05$). However, the forearm venous pressure-volume curve was not shifted at all (Fig. 6) [4]. The stiffness constant, which was calculated with a good fit to an exponential curve ($r = 0.97 ± 0.01$), was slightly elevated, although insignificantly (1.29 ± 0.2 to 1.61 ± 0.4; NS) (Fig. 7) [4].

Discussion

Venous Stiffness in Congestive Heart Failure

Elevation of venous stiffness in patients with heart failure has been repeatedly reported [1,2]. Several underlying mechanisms have been proposed [6], including elevated sympathetic nervous tone, increases in circulating catecholamines and other vasoactive substances, and local factors such as tissue pressures, interstitial edema, and organic changes in the venous vessel wall.

However, there have been few reports on the quantitative relationships between venous stiffness and clinical and hemodynamic severity in patients with congestive heart failure. In this study, the venous stiffness constant k and compliance at a venous pressure of 20 mmHg (V20) were found to be closely related to the severity of congestive heart failure from the standpoints of NYHA classification and hemodynamics. The relationship between the two parameters was not linear, and k was related much more intimately to other hemodynamic indexes. The superiority of k is understandable in that k is determined by the whole shape of the venous pressure-volume curves, while V20 is determined by only one selected position on the curve.

Plethysmography

Strain gauge arch plethysmography is a very useful method to measure peripheral circulation noninvasively [7,8]. It does have its shortcomings and problems. It measures only changes from initial values, not absolute values. Justification of volume measurements depends on the assumption that the shape of the cross section of the measured portion (forearm in this study) remains unchanged throughout the procedures and that changes in the girth length are small enough. Despite these restrictions, the method is most frequently used in clinical studies because of its practicability and reliability. The significance and reliability of measuring the forearm vein to assess a patient's venous stiffness as a whole, is another problem. The difficulty in measuring venous hemodynamics in central parts, and the absence of reports clarifying the distribution of venous wall mechanical characteristics throughout the body, seem to make plethysmographic measurement on extremities the method of practical choice.

Effects of Drugs

Clear findings consistent with the expected effect of nitroglycerin [9,10] can be interpreted as supporting evidence for the reliability and sensitivity of the method used in this study. In contrast, the effect obtained with alpha human natriuretic polypeptide is intriguing. In contrast to expectations brought about by its effect on cardiac filling pressures [11], we could not find any clear evidence of its venodilating action. Thus we conclude that hANP is much less effective at evoking venodilatation than its other mechanical actions in patients with congestive heart failure [12].

Summary and Conclusion

Venous distensibility was quantified by the forearm venous stiffness constant, which turned out to be clearly related to the severity of clinical symptoms, hemodynamic changes, and plasma concentrations of norepinephrine, in patients with congestive heart failure. The index may play an important role in understanding and evaluating congestive heart failure and the effects of drugs.

Acknowledgment. We express our sincere gratitude to the house staff and attending physicians of the 2nd Department of Internal Medicine, University of Tokyo for their assistance.

References

1. Wood JE, Litter J, Wilkins RW (1956) Peripheral vasoconstriction in human congestive heart failure. Circulation 13:524–527
2. Zelis R, Flaim SF (1982) Alterations in vasomotor tone in congestive heart failure Progr Cardiovasc Dis 24:437–459
3. Cohn JN, Franciosa JA (1977) Vasodilator therapy of cardiac failure. New Engl J Med 297:27–31
4. Ikenouchi H, Sato H, Hirata Y, Iizuka M, Serizawa T, Sugimoto T (1989) Lack of venodilative activity of alpha human atrial natriuretic polypeptide in patients with congestive heart failure. Jpn Heart J 30:443–457
5. Ikenouchi H, Iizuka M, Sato H, Momomura S, Serizawa T, Sugimoto T (1991) Forearm venous distensibility in relation to severity of symptoms and hemodynamic data in patients with congestive heart failure. Jpn Heart J 32:17–34
6. Zelis R (1974) The contribution of local factors to the elevated venous tone of congestive heart failure. J Clin Invest 54:219–224
7. Whitney RJ (1953) The measurement of volume changes in human limbs. J Physiol 121:1–27
8. Wood JE, Eckstein JW (1958) A tandem plethysmograph for study of acute responses. J Clin Invest 37:41–50
9. Mason DT, Braunwald E (1965) The effects of nitroglycerin and amyl nitrite on arteriolar and venous tone in human forearm. Circulation 32:755–766
10. Zelis R, Mason DT (1975) Isosorbide dinitrate: Effect on the vasodilator responses of nitroglycerin. JAMA 234:166–170
11. Saito Y, Nakao K, Nishimura K, Sugawara A, Okumura K, Obata K, Sonoda R, Ban T, Yasue H, Imura H (1987) Clinical application of atrial natriuretic polypeptide in patients with congestive heart failure. Circulation 76:115–124
12. Tripoddo NC, Cole FE, Frohlich ED, MacPhee A (1986) Atrial natriuretic peptide decreases circulatory capacitance in areflexic rats. Circ Res 59:291–296

Baroreflex Modifies the Effect of Vasodilators on Systemic Capacitance Vessel in Dogs

Hiroyasu Ito, Shinya Minatoguchi, Kiyoji Asano, Hisayasu Wada, Kuniyuki Takai, Masatoshi Koshiji, Yoshihiro Uno, Tomonori Segawa, Kiyoaki Inoue, and Senri Hirakawa[1]

Abstract. We examined whether the vasodilating effect of vasodilators on systemic capacitance vessels is modified by the baroreflex when a fall in arterial blood pressure was induced by the vasodilators in dogs. We estimated the dilation of the systemic capacitance vessels from the fall in mean circulatory pressure (MCP), proposed by Guyton et al. and the dilation of the systemic capacitance vessels from the fall in the total peripheral resistance (TPR). The dose-response curve for percent change in TPR to graded doses of nitroglycerin (NG, 0.8–200 µg/kg) in the untreated group was not different from that in the total spinal anesthesia (TSA) group in which the baroreflex was eliminated. The TSA group gave a dose-response curve for percent change in MCP to NG on the low dose side of NG as compared with the untreated group. NG increased both plasma levels of nor-epinephrine (NA) and epinephrine (A) in the untreated group. However, NG hardly affected plasma levels of NA or A in the TSA group. Milrinone significantly decreased the TPR and there was no significant difference in the percent change in TPR between the untreated and TSA groups. Milrinone did not change the MCP in the untreated group, but decreased it in both TSA group and the group pretreated with α-blocker, suggesting that baroreflex-mediated vasoconstriction is mediated through α-adrenoceptors in the capacitance vessels. Milrinone also increased plasma levels of NA and A in the untreated group but not in the TSA group.

These results suggest that the administration of vasodilators, which induces strong and rapid dilation of systemic resistance vessels, causes the baroreflex, and that the vasodilating effect of vasodilators on the systemic capacitance vessels is strongly modified by this baroreflex-mediated venoconstriction.

Key words: Baroreflex—Mean circulatory pressure—Systemic capacitance vessels—Vasodilators—Total spinal anesthesia

---

[1] Second Department of Internal Medicine, Gifu University School of Medicine, 40 Tsukasa-machi, Gifu, 500 Japan

Introduction

It is well known that a fall in blood pressure elicits baroreflex, which causes vasoconstriction of systemic resistance vessels to restore the blood pressure back to near normal. Under these circumstances, it is easy to speculate that the systemic capacitance vessels constrict and the venous return increases, leading to an increase in cardiac output so that the blood pressure is maintained at a near normal level. However, the behavior of systemic capacitance vessels in response to baroreflex caused by the fall in blood pressure is still unknown because of the difficulties in assessing the constriction and dilation of the systemic veins.

Effects of drugs on veins are often assessed in excised veins. However, the responses to drugs differ between sites of excised veins [1–3]. It is inappropriate to extrapolate the results obtained from veins excised from a specific site to the overall responses of the systemic capacitance vessels. To assess the effect of drugs on systemic veins, occlusion plethysmography in the forearm [4,5] is sometimes used in human studies and the cardiopulmonary bypass (extracorporeal reservoir method) is used in animal studies [6,7], to measure the changes of blood volume in the reservoir. Recently, the effects of drugs on systemic veins have been evaluated by the cardiopulmonary bypass method using extracorporeal circulation apparatus during cardiac surgery [8]. However, this method has a deficiency in that the capacitance vessels may be misjudged to have constricted when the blood volume in the reservoir increases, where the true reason is the arteriolar dilation.

Guyton et al. proposed the concept of mean circulatory pressure (MCP) [9]. We evaluated whether the capacitance vessels were dilated or constricted by measuring the MCP in anesthetized open-chest dogs [10]. The aim of the present study was to examine whether the vasodilating effect of vasodilators on systemic capacitance vessels is modified by the baroreflex when a fall in arterial blood pressure is induced by the vasodilators.

Measurements of the Mean Circulatory (Filling) Pressure (MCP) in Anesthetized Dogs

As shown in Fig. 1, in open-chest dogs subjected to intravenous pentobarbital anesthesia and artificial ventilation, a Cournand catheter was inserted into the aorta through the left carotid artery and another Cournand catheter was inserted through the right jugular vein into the right atrium, through which the mean blood pressure (MBP) and mean right atrial pressure (RAP) were measured, respectively. Cardiac output (CO) was measured with an electromagnetic flow probe which was placed around the root of the aorta. Two arteriovenous shunts were produced by connecting the bilateral femoral arteries and veins with tubing and placing a constant-flow pump midway in these shunts to translocate blood rapidly, when necessary, from arteries to veins. Electrodes

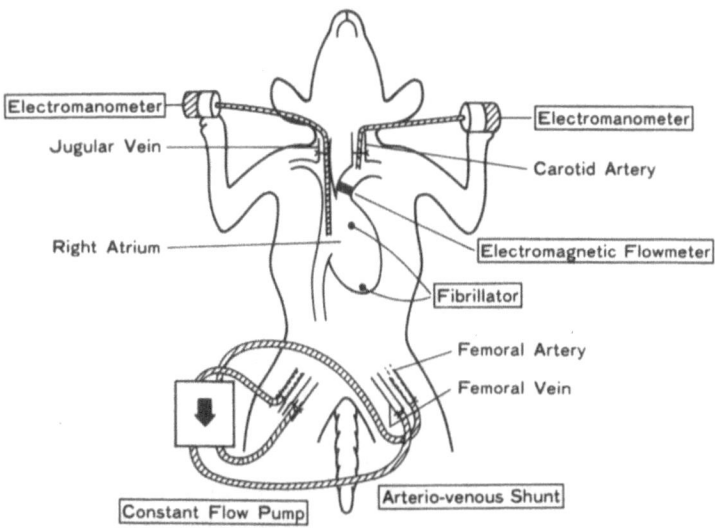

FIG. 1. Experimental arrangements for measuring mean circulatory pressure (*MCP*)

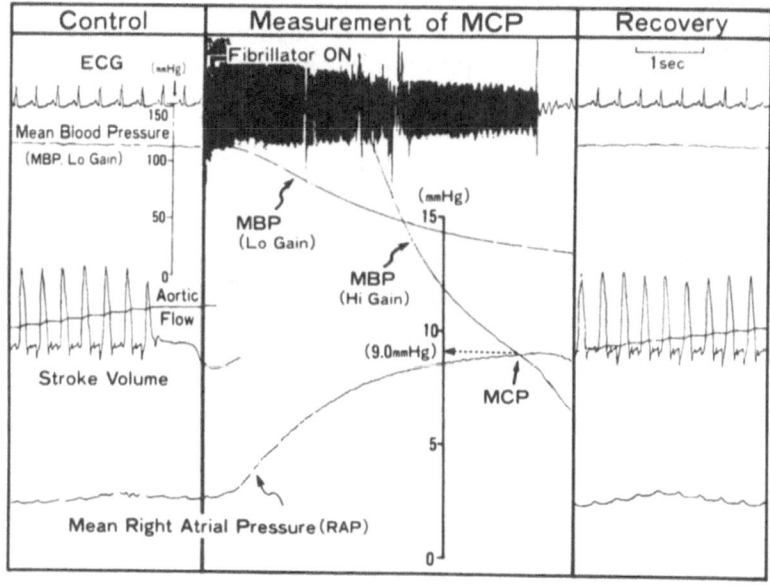

FIG. 2. A typical record of the measurement of mean circulatory pressure (*MCP*). *MBP*, mean blood pressure; *Lo*, low; *Hi*, high

were placed on the right auricle and at the apex of the heart so that ventricular fibrillation could be induced with minimal delay, when necessary, by applying alternate current (20 V, 10 Hz).

Figure 2 shows a typical record of the measurement of MCP. Ventricular fibrillation was induced at the time point marked "Fibrillator ON" in Fig. 2 by applying alternate current. Then the arteriovenous shunts were opened. Blood was translocated rapidly from the arteries to the veins by operating the constant flow pump. The MBP (high gain) fell and RAP rose, and they reached an equilibrium within 5 s. This equilibrium pressure corresponds to MCP. The MCP was 9 mmHg in this case. Immediately after determining the MCP, the heart was defibrillated by using the defibrillator (30 W-s). Hemodynamic parameters about 15 min after defibrillation at the start of the "Recovery" phase in Fig. 2 were almost restored to the control values before ventricular fibrillation.

The first measurement of MCP was performed about 30 min after completion of instrumentation. The second measurement of MCP was performed after the drugs had been administered intravenously about 15 min after the first measurement of MCP, when mean blood pressure was almost restored to the control value.

To eliminate the baroreflex, total spinal anesthesia (TSA) was performed by injecting bupivacaine (5 mg/kg) into the subarachnoid space after the spinal fluid was extracted. A small dose of epinephrine was infused continuously to maintain the mean blood pressure at approximately 70 mmHg (TSA 70) or 100 mmHg (TSA 100). When the mean blood pressure was maintained at approximately 100 mmHg, the value of the MCP was about 12 mmHg. This value was higher than that in the untreated group which normally had an MCP of about 9 mmHg. In order to match the value of MCP in the TSA group to that in the untreated group, the mean blood pressure was maintained at approximately 70 mmHg. Under this condition, the MCP was about 9 mmHg.

As shown in Table 1, the reproducibility of MCP and TPR was good.

All values were expressed as mean ± SE. Comparisons between two groups were assessed by the Student's t-test. Comparisons between three or more groups were assessed by analysis of variance followed by the Bonferroni method. A P value less than 0.05 was considered to be statistically significant.

Evaluation of Baroreflex-Induced Venoconstriction

Figure 3 shows the changes in hemodynamic parameters such as MCP in response to intravenous administration of 50 µg/kg of nitroglycerin (NG) in the untreated group (upper panel) and in the TSA 100 group (lower panel). Epinephrine was continuously infused to maintain a "normal" blood pressure in the TSA 100 group. When we compared the untreated and TSA 100 groups, there was no difference in the control value of mean blood pressure (untreated group: 93.8 ± 4.5 mmHg, TSA 100 group: 102.0 ± 5.0 mmHg). However, the control value of MCP was significantly greater in the TSA 100 group (12.3 ±

TABLE 1. The reproducibility of the measurement of mean circulatory pressure (MCP) and total peripheral resistance (TPR).

	MCP (mmHg)		TPR (dyne·s·cm^{-5})	
	1st	2nd	1st	2nd
Untreated group ($n = 5$)	8.4 ± 0.5	8.2 ± 0.4	7539 ± 380	7035 ± 333
TSA group ($n = 6$)	10.0 ± 0.5	10.0 ± 0.5	5969 ± 426	5929 ± 191

TSA, total spinal anesthesia.

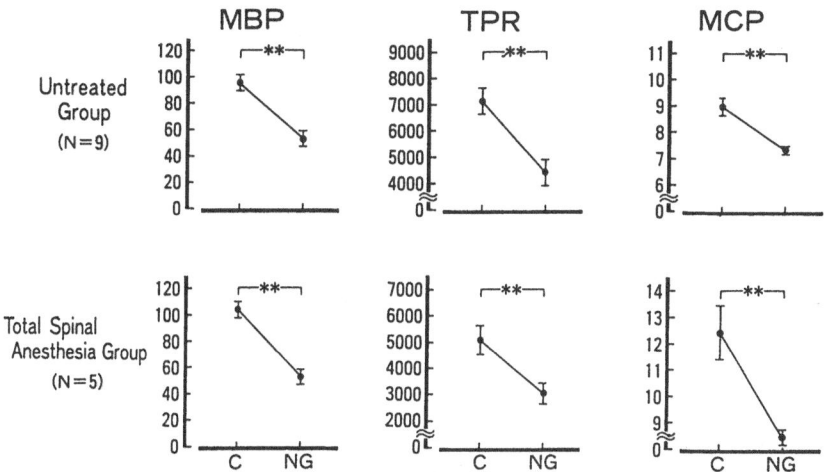

FIG. 3. Effects of intravenous administration of nitroglycerin (*NG*) on mean blood pressure (*MBP*), total peripheral resistance (*TPR*), and mean circulatory pressure (MCP) [mean ± SE]. The *upper panels* show untreated dogs, and the *lower panels*, dogs under total spinal anesthesia. *C*, control; *NG*, nitroglycerin (50 µg/kg), **$P < 0.01$

1.0 mmHg) than in the untreated group (8.9 ± 0.3 mmHg). Under these circumstances, NG decreased both TPR and MCP, suggesting that NG dilates both systemic resistance vessels and systemic capacitance vessels. There was no significant difference in the percent changes of TPR in response to NG between the two groups (untreated group: −39.1% ± 3.2%, TSA 100 group: −39.9% ± 3.5%). However, the percent change of MCP in response to NG was significantly greater in the TSA group (−30.2% ± 4.5%) than in the untreated group (−17.1% ± 2.0%). The explanation of these results may be that (1) the initial value of MCP was greater in the TSA 100 group than in the untreated group or (2) the venodilating effect of NG on the systemic capacitance vessels was partly diminished by the baroreflex-mediated venoconstriction in the untreated group.

Figure 4 shows the dose-response curves for the percent change of TPR (left panel) and the percent change of MCP (right panel) to graded doses of NG in

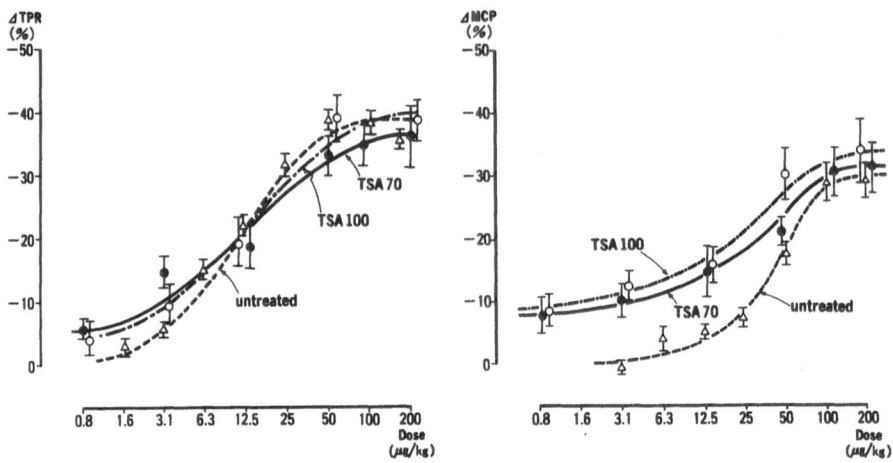

FIG. 4. Dose-response curves for percent change in total peripheral resistance (% Δ TPR) and in mean circulatory pressure (% Δ MCP) to graded doses of nitroglycerin (NG) in the untreated group and TSA 100 and TSA 70 groups (mean ± SE). *TSA*, total spinal anesthesia; *TSA 100*, TSA group in which mean blood pressure was maintained at about 100 mmHg; *TSA 70*, TSA group in which mean blood pressure was maintained at about 70 mmHg

the untreated group (dashed curve), TSA 100 (long dashed curve), and TSA 70 groups (solid curve), in which the mean blood pressure was maintained at about 100 mmHg and 70 mmHg, respectively, by continuous intravenous infusion of epinephrine. The dose-response curve for the percent change of TPR to NG in the untreated group was not different from that in the TSA 100 group or in the TSA 70 group. Both the TSA 100 group and TSA 70 group gave a dose-response curve for the percent change of MCP to NG on the left side (low-dose side of NG) as compared with the untreated group, suggesting that the vasodilating effect of NG on the capacitance vessels is potentiated by eliminating the baroreflex.

These results suggest that the vasodilating effect of NG on the capacitance vessels in the baroreflex-eliminated group is greater than in the untreated group, regardless of the initial value of MCP.

Figure 5 shows the effect of milrinone (100 µg/kg) in terms of the percent change of TPR and the percent change of MCP in the untreated group, in the TSA 100 group, and in the group pretreated with α-blocker (phenoxy-benzamine). In all these three groups, milrinone significantly decreased both the mean blood pressure (percent change of MBP, untreated group: -24.4% ± 2.4%; TSA 100 group: -21.9% ± 1.8%; α-blocker-pretreated group: -20.7% ± 2.6%) and the TPR (percent change of TPR, untreated group: -34.3% ± 1.3%; TSA 100 group: -30.4% ± 2.5%; α-blocker pretreated group: -36.3% ± 2.1%). There was no significant difference in the percent change of TPR

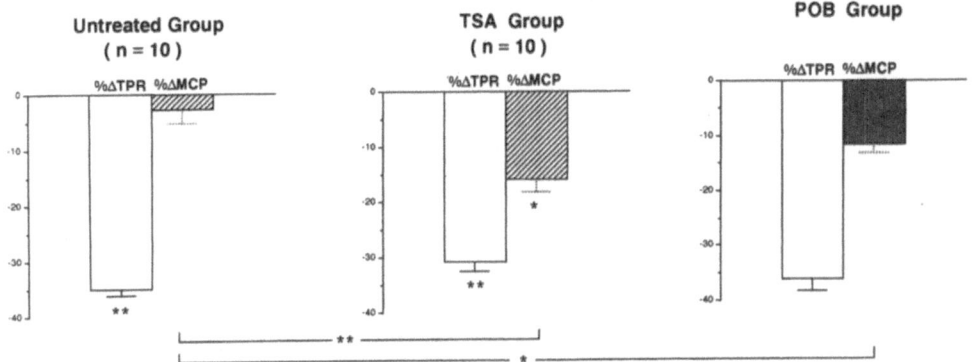

FIG. 5. Effects of milrinone (100 μg/kg) on percent changes in total peripheral resistance (% Δ *TPR*) and mean circulatory pressure (% Δ *MCP*) [mean ± SE]. *TSA*, total spinal anesthesia; *TSA group*, group in which mean blood pressure was maintained at about 100 mmHg; *POB*, phenoxybenzamine; *$P < 0.05$; **$P < 0.01$

among these three groups. On the other hand, milrinone did not change the MCP in the untreated group, but decreased it in both the TSA 100 group (−16.3% ± 2.2%) and the α-blocker pretreated group (−11.7% ± 1.6%). There was no significant difference in the degree of decrease in MCP between the latter two groups. When the baroreflex was eliminated, the vasodilating effect of milrinone on the capacitance vessels was potentiated. This again suggests that the vasodilating effect of milrinone on the capacitance vessels in the untreated group is minimized by baroreflex-mediated vasoconstriction. The vasodilating effect of milrinone on the capacitance vessels was potentiated also in the α-blocker pretreated group, suggesting that baroreflex-mediated vasoconstriction is mediated through α-adrenoceptors in the capacitance vessels.

Figure 6 shows the percent change in mean blood pressure and plasma levels of norepinephire (NA) and epinephrine (A) due to the intravenous injection of NG. The doses used were 12.5 and 100 μg/kg in the untreated group and 100 μg/kg in the TSA 100 group. In the untreated group, NG decreased the mean blood pressure (12.5 μg/kg: −13.4% ± 2.5%; 100 μg/kg: −42.3% ± 5.4%) in a dose dependent manner, and increased the plasma levels of norepinephrine (12.5 μg/kg: 94.0% ± 8.2%, 100 μg/kg: 156.4% ± 11.5%) and epinephrine (12.5 μg/kg: 122.8% ± 10.8%, 100 μg/kg: 177.6% ± 20.4%) in a dose-dependent manner. Especially, the percent increase in plasma epinephrine level due to NG administration at 12.5 μg/kg was significantly greater than that in the plasma norepinephrine level. However, in the TSA 100 group, NG hardly affected the plasma levels of norepinephrine (6.1% ± 3.8%) or epinephrine (1.1% ± 2.1%) although mean blood pressure (−39.6% ± 3.1%) fell to the same degree as in the untreated group, suggesting that baroreflex was completely abolished in the TSA 100 group.

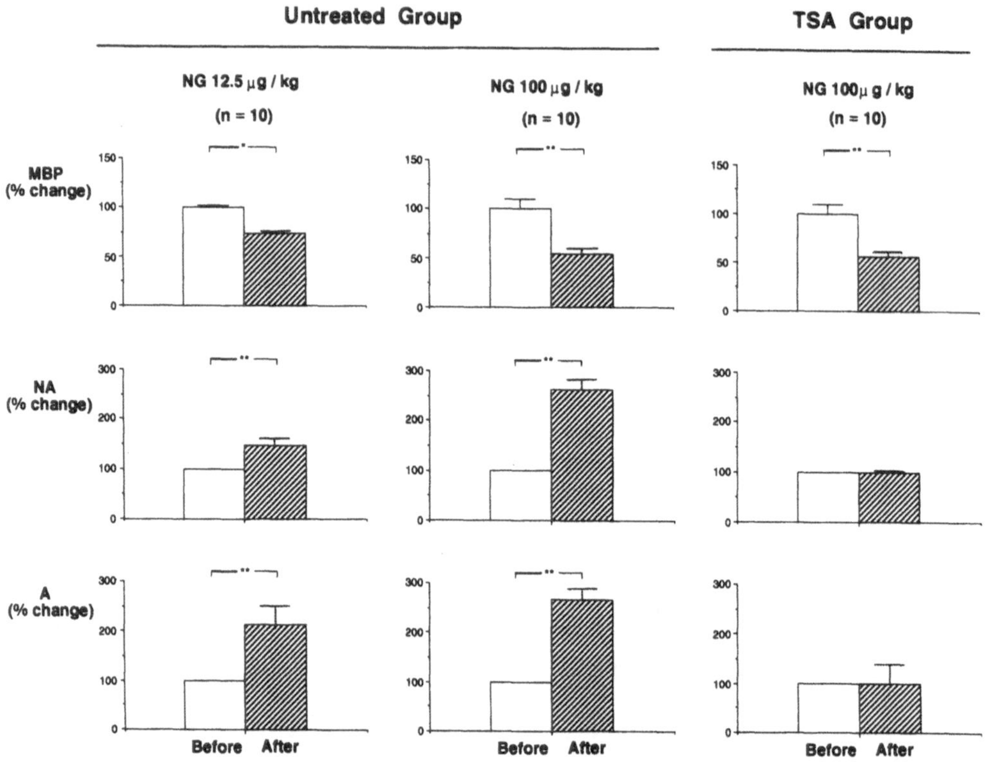

Untreated Group   **TSA Group**

NG 12.5 µg / kg   NG 100 µg / kg   NG 100µ g / kg

(n = 10)   (n = 10)   (n = 10)

MBP
(% change)

NA
(% change)

A
(% change)

Before After   Before After   Before After

FIG. 6. Effects of intravenous administration of nitroglycerin (*NG*) on mean blood pressure (*MBP*), plasma norepinephrine (NA) and epinephrine (A) concentrations in the untreated group and the TSA group (mean ± SE). *TSA group*, group in which mean blood pressure was maintained at about 100 mmHg; *TSA*, total spinal anesthesia; *P < 0.05; **P < 0.01

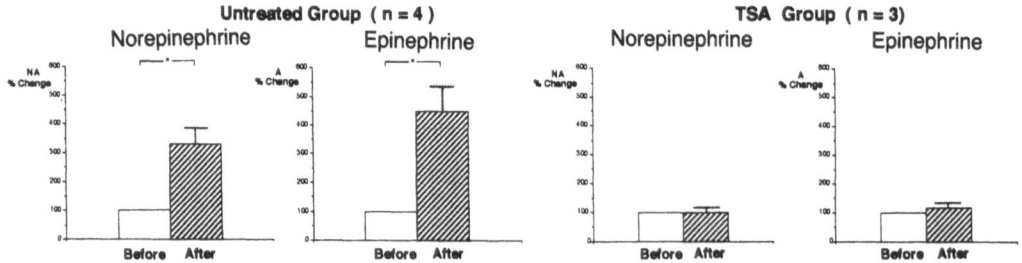

Untreated Group (n = 4)   **TSA Group (n = 3)**
Norepinephrine   Epinephrine   Norepinephrine   Epinephrine

Before After   Before After   Before After   Before After

FIG. 7. Effects of milrinone (100 µg/kg) on percent changes in plasma norepinephrine (NA) and epinephrine (A) concentrations in the untreated group and the TSA group (mean ± SE). *TSA*, total spinal anesthesia; *TSA group*, group in which mean blood pressure was maintained at about 100 mmHg; *P < 0.05; **P < 0.01

Figure 7 shows the percent change in plasma levels of norepinephrine and epinephrine due to the intravenous injection of milrinone. In the untreated group, milrinone decreased the mean blood pressure (103.0 ± 2.6 → 78.5 ± 3.4 mmHg, −24.4% ± 2.2%) and increased the plasma levels of norepinephrine (221.1% ± 51.1%) and epinephrine (346.2% ± 98.2%). The percent increase in plasma epinephrine level was greater than that in plasma norepinephrine level in the untreated group. However, in the TSA 100 group, milrinone hardly affected plasma levels of norepinephrine or epinephrine, although mean blood pressure (100.0 ± 1.2 → 76.7 ± 1.8 mmHg, −23.4% ± 0.9%) fell to the same degree as in the untreated group.

Possible Mechanisms for Modification by Baroreflex-Mediated Venoconstriction of the Effect of Vasodilators on Systemic Capacitance Vessels

The vasodilating effect of vasodilators on the systemic capacitance vessels was clearly modified by baroreflex-mediated venoconstriction. However, the effect of vasodilators on the systemic resistance vessels was not modified by baroreflex. One possible explanation can be advanced for this difference by considering the changes in plasma catecholamine concentrations in the untreated group and TSA group (baroreflex-eliminated group). The order of importance of adrenoceptor type for the stimulatory action of norepinephrine is said to be α-adrenoceptors > β-adrenoceptors, whereas for epirephrine, β-adrenoceptors > α-adrenoceptors [11]. Both α-adrenoceptors and β-adrenoceptors exist in the systemic resistance vessels. The stimulation of α-adrenoceptors causes the vasoconstriction and the stimulation of β-adrenoceptors causes the vasodilation in the systemic resistance vessels. However, it is reported that venoconstriction via the stimulation of α-adrenoceptors overwhelms the β-adrenoceptor-mediated venodilation [12]. According to our study, when mean blood pressure fell by about 30 mmHg after the administration of vasodilators, plasma norepinephrine concentration increased and at the same time plasma epinephrine concentration also increased markedly (Figs. 6, 7). This increase in plasma epinephrine concentration seems to be due to the increase in the release of epinephrine from the adrenal medulla because of the severe, near-shock hypotension induced by vasodilators.

The left panel of Fig. 8 shows the difference in the constriction and dilation between the systemic resistance vessels and systemic capacitance vessels. In the systemic resistance vessels, epinephrine elicits β-adrenoceptor-mediated vaso-dilation as well as α-adrenoceptor-mediated vasoconstriction. The constriction of systemic resistance vessels via the stimulation of α-adrenoceptors by epinephrine was counterbalanced by the dilation of systemic resistance vessels via the stimulation of β-adrenoceptors. As a result, the systemic resistance vessels appear unchanged. However, in the capacitance vessels, constriction

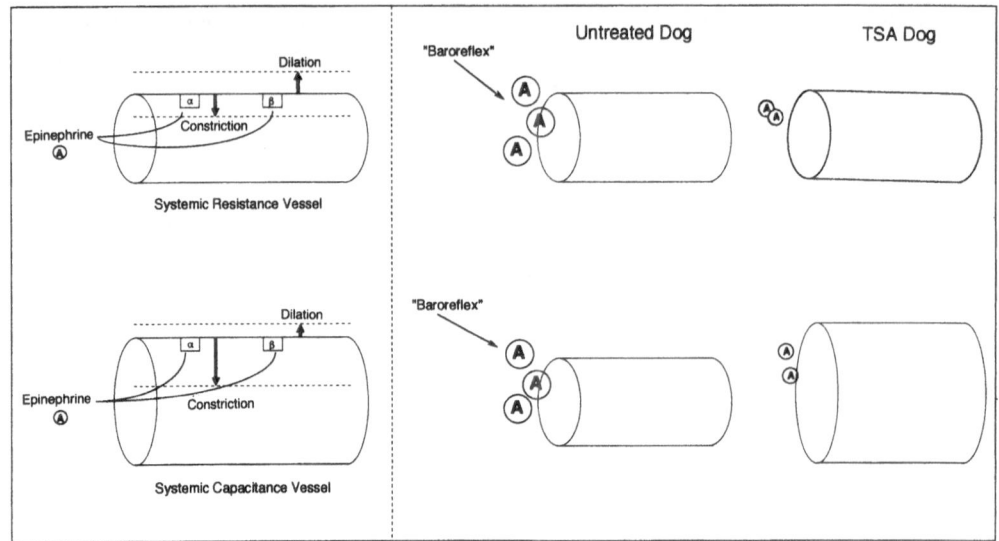

Fig. 8. A schematic explanation for the mechanism by which the vasodilating effect of vasodilators on systemic capacitance vessels is modified by the baroreflex. The *upper diagrams* represent a systemic resistance vessel, and the *lower diagrams*, a systemic capacitance vessel. The *left panel* shows epinephrine (*A*) effects mediated through adrenoceptors, and the *right panel* indicates the effect of baroreflex or "shock". *TSA*, total spinal anesthesia; α, β, α- and β-adrenergic receptors

of systemic capacitance vessels via the stimulation of α-adrenoceptors by epinephrine may overwhelm the dilation of systemic capacitance vessels via the stimulation of β-adrenoceptors. As a result, the capacitance vessels appear constricted.

The right panel of Fig. 8 shows the constriction and dilation of systemic resistance vessels and systemic capacitance vessels in the presence of increased plasma epinephrine induced by baroreflex or "shock" in the untreated and the TSA (baroreflex eliminated) groups. In the untreated group, although the systemic resistance vessels are not modified by epinephrine, the systemic capacitance vessels are modified by the constricting effect of baroreflex. In the TSA group, the systemic capacitance vessels do not constrict, judging from the fact that the plasma epinephrine level was hardly elevated after the administration of vasodilators (Figs. 6, 7). These results support the concept that the vasodilating effect of vasodilators on the systemic capacitance vessels is fairly strongly modified by epinephrine-mediated venoconstriction, while the effect on the systemic resistance vessels is hardly modified by epinephrine.

References

1. Folkow B, Mellander S (1964) Veins and venous control. Am Heart J 68:397–408
2. Rothe CF (1983) Venous system: Physiology of the capacitance vessels. In: Handbook of physiology, sect 2, vol 3. American Physiological Society, Bethesda, pp 397–452
3. Bocking JK, Roach MR (1974) Effects of hydrostatic pressure on the elastic properties of cat veins. Can J Physiol Pharmacol 52:149–152
4. Burch GE (1947) A new sensitive portable plethysmograph. Am Heart J 33:48–75
5. Mason DT, Baunwald E (1965) The effects of nitroglycerin and amyl nitrate on arterior and venous tone in the human forearm. Circulation 32:755–766
6. Caldini P, Permutt S, Waddell JA, Riley RL (1974) Effect of epinephrine on pressure, flow, and volume relationships in the systemic circulation of dogs. Circ Res 34:606–623
7. Tanaka Y, Morimoto T, Miki K, Nose H, Miyazaki M (1981) On-line control of circulating blood volume. Jpn J Physiol 31:427–431
8. Mario RJ, Romagnioli A, Keats A (1975) Selective venoconstriction by dopamine in comparison with isoproterenol and phenylephrine. Anesthesiology 43:570–572
9. Guyton AC, Jones CE, Coleman TG (1973) Regulation of venous return. In: Circulatory physiology: Cardiac output and its regulation. Saunders, Philadelphia, pp 173–236
10. Ito H, Hirakawa S (1984) Symposium on heart failure and vasodilator therapy: Effect of vasodilators on the systemic capacitance vessels. A study with the measurement of the mean circulatory pressure in dogs. Jpn Circ J 48:388–404
11. Hoffman BB, Lefkowitz RJ (1990) Catecholamines and sympathomimetic drugs. In: Gilman AG (ed) The pharmacological basis of therapeutics, 8th edn. Pergamon, New York, pp 221–243
12. Hirakawa S, Ito H, Kotoo Y, Abe C, Endo T, Takada N, Fuseno H (1984) The role of alpha and beta adrenergic receptors in constriction and dilation of the systemic capacitance vessels; a study with measurements of the mean circulatory pressure in dogs. Jpn Circ J 48:620–632

Regulation of Hepatic Vascular Capacitance

CARL F. ROTHE[1]

Abstract. The mammalian liver is part of the splanchnic bed that provides a highly significant blood reservoir for compensation for loss of blood from the thorax and consequently a decreased cardiac output. Stresses requiring blood volume compensation include: blood or water loss, quiet standing, return from a prolonged bout of zero gravity, thermal stress, or exercise. Hepatic blood volume is changed *passively* by changes in hepatic arterial or gastrointestinal blood flow or by changes in vena caval pressure. Liver blood volume is also changed *actively* via reflexes, hormones, or drugs that activate the hepatic venous smooth muscle. Hepatic vascular *compliance* is about ten times that of the body as a whole, and is changed by active mechanisms. The hepatic *unstressed blood volume* is markedly changed by hormones. *Baroreceptor* control of hepatic volume is important in dogs, but not in cats and possibly not in humans. The hepatic venular pressure of rats, rabbits, and puppies is about 40% of the pressure gradient between the portal vein and vena cava. Major areas of uncertainty include: (a) the magnitude of sinusoidal pressure and how it is controlled, (b) the mechanisms controlling hepatic venous resistance, and (c) the mechanisms influencing hepatic vascular capacitance.

Key words: Liver—Vascular capacitance—Hepatic blood volume

Introductian

The liver, besides its crucial metabolic function, may provide a blood reservoir function. Changes in vascular capacitance and subsequent blood volume redistribution provide compensation for changes in effective blood volume. An adequate blood volume is needed for right ventricular filling and thus an adequate cardiac output. Examples of stresses requiring compensation include hemorrhage or water loss [1], the effect of standing, in which blood and fluid

[1] Department of Physiology and Biophysics, Indiana University School of Medicine, 635 Barnhill Drive, Indianapolis, IN 46202–5120, USA

are pooled in the legs [2,3], thermal stress, in which there is both loss of water from perspiration and accumulation of blood in the hot skin [4], and exercise. Because the splanchnic bed contains about one-third of the total blood volume, it can act as a major reservoir for the needed blood volume redistribution under conditions of stress [5,6]. Because the liver has such a large blood content (about 35% under normal conditions) and provides a large part of the splanchnic bed (2.5% of the body weight), it is potentially a major component of this reservoir. The many papers of Greenway and colleagues, e.g. [6], provide an excellent source of information about the circulation of the liver.

Because there are uncertainties and differences of opinion with respect to terminology, a few definitions are in order [7–9]. *Vascular capacitance* relates the volume of blood contained within the vasculature to the ·pressure distending the vascular walls. Vascular *compliance* is the ratio of the change in contained volume in response to a change in distending pressure. The *stressed volume* is that blood volume contained in a vascular segment because of the distending pressure. *Unstressed volume* is the volume contained in a vascular segment at a distending pressure of zero, and is obtained by extrapolating the pressure-to-volume relationship over the normal operating range to zero. *Vascular capacity* is the total blood volume contained at a specific distending pressure, and is the sum of the stressed and unstressed volumes.

There are two major mechanisms—passive and active—by which blood volume can be redistributed to or from a vascular bed [7]. A *passive* mechanism involves elastic distension or recoil of the vasculature. A change in venous (outflow) pressure can act to change the distending pressure and thus the volume. Furthermore, because venous outflow resistance is always finite, a change in flow across this outflow resistance will lead to a change in the pressure gradient. This will change the upstream distending pressure if the outflow venous pressure is unchanged and, consequently, will cause a change in volume. A decrease in flow leads to a decrease in volume of blood stored in an organ, whereas an increase of blood flow leads to an increase in volume.

An *active response* occurs when the smooth muscle around the veins contracts or relaxes and thus acts to change the distending pressure or volume, or both. Sympathetic nerve activity can cause both *veno*constriction (a decrease in venous volume from contraction of the smooth muscle) and *vaso*constriction (an increase in vascular resistance from contraction of the smooth muscle surrounding arterioles). One complication is that the reduction in venous volume also adds resistance. This attenuates, but does not fully negate, the decrease in volume. This occurs because the added resistance leads to an increased pressure gradient and so upstream distending pressure. There are important differences in response between hepatic nerve stimulation, which acts only on the vessels of the liver, and splanchnic nerve stimulation, which also acts to decrease portal venous flow. To distinguish between active and passive changes due to reflex activity, ganglionic blocking agents can be used. Both hexamethonium and atropine must be used if the neural influences on the liver are to be adequately blocked [10]. Understanding the neural and hormonal control of the liver circulation continues to be uncertain, because of differences in anatomy of the liver between species and its complex structure [11,12].

Not only does the liver contain a large fraction of the blood, but the blood flow through the liver per kilogram of tissue is about 10 times that of the body as a whole [6]. Thus, about 25% of the cardiac output passes through the liver. The compliance of the vasculature of the liver is $20-30\,\text{ml} \cdot \text{mmHg}^{-1} \cdot \text{kg}^{-1}$ liver [6,13,14]. This is also about 10 times of that of the body as a whole.

In studies completed several years ago [13,14], Tom Bennett and I vascularly isolated the liver by cannulating the portal vein, diverting the flow to a reservoir, and then pumping blood to the liver at a known constant rate via the downstream part of the portal vein. We also cannulated the hepatic artery and interposed a pump for constant hepatic arterial perfusion. In addition, by placing a cannula in the vena cava with appropriate bypasses, we collected all of the blood coming from the liver and pumped it to a jugular vein. Any excess of blood in the reservoir was pumped back to the dog through the jugular veins. The caval cannula had a coaxially-placed catheter to sense hepatic outflow pressure so that we could, with automatic servo-control, rapidly adjust the pump rate to maintain the hepatic venous pressure at a fixed, constant level [14].

With this technique and integration of the two inflows minus the outflow to determine changes in volume, we studied the effect of changes in blood flow on the liver blood volume. We found that liver volume changed about 0.066 ml per ml/min change in flow, a value similar to that of other tissues [14-16]. (For the splanchnic bed as a whole, Brooksby and Donald [17] reported a value of 0.19 ml per ml/min flow change.) The relationship is almost linear. There was no significant difference on the magnitude of blood volume change between changing hepatic arterial or portal venous flows. We therefore concluded that most of the vascular compliance is downstream from the junction of the hepatic artery and portal vein. What we do not know is whether the volume change is primarily at the level of the sinusoids or the intrahepatic veins, however.

In that study, we also confirmed that with a constant inflow rate, as the hepatic venous pressure is changed between zero and about 3 mmHg the portal venous pressure does not appreciably change [14,18]. However, at hepatic venous pressures above about 3.5 mmHg, the increase in pressure in the portal vein parallels an increase in hepatic venous pressure. We also plotted portal venous pressure at the various flows and found that it could be fitted by a straight line, but the intercept of the line at zero flow was at about 2.5 mmHg. Thus, the resistance of the hepatic outflow veins is not constant.

Three hypotheses have been invoked to explain the lack of full pressure transmission upstream: (1) A sphincter acting like a vascular waterfall [18,19]. (2) A pressure-sensitive outflow resistance [20,21]. (As the venous pressure is increased, the veins distend. The increased radius of the outflow veins consequently decreases the resistance to flow, as predicted by the Poiseuille equation.) (3) Hepatic interstitial pressure that partially collapses the veins (a

Starling resistor). To choose between these hypotheses, the interlobular hepatic vein, venule, sinusoidal, and interstitial pressures of the liver must be measured accurately.

We have also reported [13,14] that the vascular compliance of the dog liver is $20-25\,\text{ml}\cdot\text{mmHg}^{-1}\cdot\text{kg}^{-1}$ tissue. The compliance is relatively linear and has a zero intercept. Greenway et al. [15] reported a value in cats of $25-30\,\text{ml}/$ (mmHg-kg). In an earlier study, they had reported a nonlinearity in the pressure-volume relationship [22]. These data suggest that most of the compliance is downstream from a nonconstant resistance.

Neural and Hormonal Control of the Liver

Hepatic nerve stimulation is potentially a potent mechanism for actively decreasing hepatic blood volume. Greenway et al. [23–26] have reported that about 50% of the blood volume can be expelled from the liver of cats and dogs. (The hepatic arterial conductance is reduced to only about 30% of control, suggesting a major passive component.) In humans the response of splanchnic blood volume may be entirely a passive response to changes in flow [4]. In dogs, we found that hepatic nerve stimulation reduced both hepatic arterial and portal venous conductance [13]. With constant perfusion, a 10 pulses/s hepatic nerve stimulation caused a 12% reduction in liver blood volume [13]. This supports the concept of the presence of an active mechanism in this species.

On stimulation of the smooth muscle of the liver with norepinephrine, Greenway et al. found a change in liver volume without an appreciable change in vascular compliance, suggesting that the unstressed volume is actively changed rather than the compliance [15]. However, while stimulating the hepatic nerves of dogs at 5 pulses/s, we reported [13] a 35% change in hepatic compliance as well as an 8% decrease in liver blood volume.

Changing carotid sinus pressure, which changes the activity of the baroreceptors, clearly changes the liver volume of dogs [27]. For example, when the carotid sinus pressure was decreased from 144 mmHg to 40 mmHg, vasoconstriction increased the arterial pressure and hepatic arterial resistance, but the change in perfusion pressure was proportionally larger than the change in resistance. Thus the hepatic arterial flows were increased. Although the liver volume changed 1.2 ml/kg body weight under these conditions (a 16% change in its volume), the increase in flow caused an increase in volume that counteracted in part the decrease in volume due to the active venoconstriction. This provides an example of the complexity of studying the liver as a volume reservoir.

Lautt et al. [28,29], using a plethysmograph to measure liver volume changes, have reported that the carotid sinus reflex has little influence in cats. With a bilateral carotid occlusion, causing a clear increase in arterial blood pressure, there was no significant change in liver volume. This response was not changed after denervation. To explore this lack of responsiveness, we attached ultrasonic crystals on both sides of a lobe of the liver and spleen of cats to estimate volume changes [30]. We isolated the carotid sinus region, perfused the brain separately, and, with constant flow perfusion, controlled the carotid sinus pressure at various known pressures. We reported [30] that not only was the arterial blood pressure increased (as a response to a decrease in carotid sinus pressure, as expected), but the cardiac output also was increased as carotid sinus pressures decreased. (The only other study of changes in cardiac output of cats during changes in carotid sinus pressure showed no significant change [31]). The spleen lobe thickness was significantly decreased and thus volume was decreased by the decreased carotid sinus pressure, as expected. Surprisingly, there was a small but significant increase in liver volume as carotid sinus pressure was decreased. Thus, the dog liver responds to the changes in carotid sinus baroreceptor activity but apparently not the cat liver.

Using technetium-99m labeled erythrocytes and a gamma camera to image the segments of the body of people, Bell et al. [32] reported that phenylephrine increased arterial blood pressure and decreased heart rate, as expected, but decreased the spleen volume and gut volume. The liver volume increased, however. A significant decrease in total splanchnic volume was the net result.

Stimulating the vasculature of the liver not only changes blood flow and liver volume. An increased hepatic venous pressure, hepatic nerve stimulation, epinephrine, and especially histamine also causes a significant increase in filtration of fluid from the hepatic vasculature [13]. This not only complicates studies of the control of hepatic blood volume, but also may be an important factor influencing hepatic metabolic function.

Problems

To understand more clearly the role of the liver as a blood volume reservoir, it is important to know what segment of the liver (portal venular, sinusoid, hepatic venule) is involved and the mechanisms by which volumes are changed. What is the magnitude of the normal hepatic sinusoidal pressure? Earlier studies [33,34] suggested that the sinusoidal pressure was within 1 mmHg of the hepatic vena caval pressure. On the other hand, Lautt and colleagues [21,35–37] have suggested that the intralobular venous pressure, and thus the sinusoidal pressure, is within 1 mmHg of the portal venous pressure. This suggests virtually zero resistance from the portal vein to the sinusoid. Because of this discrepancy, we studied the hepatic venule pressure in rats, rabbits, and puppies [38]. A servo-null micropipet system was used to measure pressure in hepatic venules that are about 25 μm in diameter at sites within 50 μm from the sinusoids. We found that the hepatic venular pressures were about 3 mmHg less than portal venous pressure and about 4 mmHg higher than caval pressure. Thus, the hepatic venular pressures are at about 40% of the pressure gradient between the portal vein and the inferior vena cava. We found that norepinephrine increased the portal venous pressure. Some norepinephrine was transmitted through the liver into the system circulation, causing an increase in arterial blood pressure. However, we found no significant change in hepatic venous pressure. Was this from technical errors? What controls the hepatic venous resistance and sinusoidal pressure? Are there mechanisms that act to control this pressure at a constant level under a variety of conditions?

In another study, we used a mechanical model, developed the theoretical equations, and after extensive surgery, carefully placed catheters at various depths of the liver to measure the pressure in the intralobular veins. From these studies [39] we conclude that: (1) there are no sphincter regions in hepatic veins >1.5 mm ID in normal dogs; (2) the major venous pressure gradient in the normal canine liver must lie upstream to the large (>2 mm ID) hepatic veins; (3) pressure measurements made with a catheter must be corrected for the resistance effect of the presence of the catheter; (4) there are no significant venous collaterals downstream from the sinusoids. (Current anatomic evidence also does not suggest the presence of venous collaterals [12,40–42]. This suggests that if hepatic microvascular pressures are to be measured, they must be measured with some transmural approach.

Because the liver is so close to the diaphragm that it moves with each breath, because the liver extracts most indicators, and because the liver has two major inflow systems and many output veins into the vena cava, definitive studies will continue to be both difficult and variable. Nonetheless, because the liver is essential for metabolism and detoxification, and because the liver is potentially such an important volume reservoir for the maintenance of cardiovascular homeostasis, further research is needed to provide a better understanding of the control of hepatic resistances and the mechanisms causing changes in hepatic blood volume.

Acknowledgment. Our research has been supported by a National Institutes of Health grant (HL07723).

References

1. Rothe CF (1983) Reflex control of veins and vascular capacitance. Physiol Rev 63:1281–1342
2. Hinghofer-Szalkay H, Moser M (1986) Fluid and protein shifts after postural changes in humans. Am J Physiol 250(1, Pt. 2):H68–H75
3. Hinghofer-Szalkay H, Kravik SE, Greenleaf JE (1988) Effect of lower-body positive pressure on postural fluid shifts in men. Eur J Appl Physiol 57:49–54
4. Rowell LB (1986) Human circulation: regulation during physical stress. Oxford University Press, New York
5. Brunner MJ, Greene AS, Frankle AE, Shoukas AA (1988) Carotid sinus baroreceptor control of splanchnic resistance and capacity. Am J Physiol 255:H1305–H1310

6. Greenway CV, Lautt WW (1989) Hepatic circulation. In: Handbook of physiology, the gastrointestinal system, motility and circulation, sect 6, vol 1, part 2, chap 41. American Physiological Society, Bethesda, MD, pp 1519–1564

7. Rothe CF (1983) Venous system: physiology of the capacitance vessels. In: Shepherd JT, Abboud FM (eds) Handbook of physiology. The cardiovascular system, sect 2, vol III, part 1. American Physiological Society, Bethesda, MD, pp 397–452

8. Rothe CF (1991) Vascular capacitance. In: Dulbecco R (ed) The encyclopedia of human biology, Vol 7. Academic Press, San Diego, pp 757–765

9. Rothe CF (1993) Mean circulatory filling pressure: Its meaning and measurement. J Appl Physiol 74:499–509

10. Greenway CV (1991) Blockade of reflex venous capacitance responses in liver and spleen by hexamethonium, atropine, and surgical section. Can J Physiol Pharmacol 69:1284–1287

11. Bauer W, Dale HH, Poulsson LT, Richards DW (1932) The control of circulation through the liver. J Physiol (Lond) 74:343–375

12. Ekataksin W, Wake K (1991) Liver units in three dimensions: I. Organization of argyrophilic connective tissue skeleton in porcine liver with particular reference to the "compound hepatic lobule". Am J Anat 191:113–153

13. Bennett TD, MacAnespie CL, Rothe CF (1982) Active hepatic capacitance responses to neural and humoral stimuli in dogs. Am J Physiol 242 (Heart Circ Physiol 11):H1000–H1009

14. Bennett TD, Rothe CF (1981) Hepatic capacitance responses to changes in flow and hepatic venous pressure in dogs. Am J Physiol 240 (Heart Circ Physiol 9):H18–H28

15. Greenway CV, Seaman KL, Innes IR (1985) Norepinephrine on venous compliance and unstressed volume in cat liver. Am J Physiol 248 (Heart Circ Physiol 17):H468–H476

16. Rothe CF, Gaddis ML (1990) Autoregulation of cardiac output by passive elastic characteristics of the vascular capacitance system. Circulation 81:360–368

17. Brooksby GA, Donald DE (1972) Release of blood from the splanchnic circulation in dogs. Circ Res 31:105–118

18. Mitzner W (1974) Hepatic outflow resistance, sinusoid pressure, and the vascular waterfall. Am J Physiol 227:513–519

19. Permutt S, Riley RL (1963) Hemodynamics of collapsible vessels with tone: The vascular waterfall. J Appl Physiol 18:924–932

20. Greenway CV, Lautt WW (1988) Distensibility of hepatic venous resistance sites and consequences on portal pressure. Am J Physiol 254 (Heart Circ Physiol 23):H452–H458

21. Lautt WW, Legare DJ, Greenway CV (1987) Effect of hepatic venous sphincter contraction on transmission of central venous pressure to lobar and portal pressure. Can J Physiol Pharmacol 65:2235–2243

22. Lautt WW, Greenway CV (1976) Hepatic venous compliance and role of liver as a blood reservoir. Am J Physiol 231:292–295

23. Greenway CV, Lautt WW (1972) Effects of infusions of catecholamines, angiotensin, vasopressin and histamine on hepatic blood volume in the anaesthetized cat. Br J Pharmacol 44:177–184

24. Greenway CV, Oshiro G (1972) Comparison of the effects of hepatic nerve stimulation on arterial flow, distribution of arterial and portal flows and blood content in the livers of anaesthetized cats and dogs. J Physiol (Lond) 227:487–501

25. Greenway CV, Dettman R, Burczynski F, Sitar DS (1986) Effects of circulating catecholamines on hepatic blood volume in anesthetized cats. Am J Physiol 250: H992–H997
26. Segstro R, Greenway CV (1986) Alpha adrenoceptor subtype mediating sympathetic mobilization of blood from the hepatic venous system in anesthetized cats. J Pharmacol Exp Ther 236:224–229
27. Carneiro JJ, Donald DE (1977) Change in liver blood flow and blood content in dogs during direct and reflex alteration of hepatic sympathetic nerve activity. Circ Res 40:150–158 ·
28. Lautt WW, Greenway CV (1972) Hepatic capacitance vessel responses to bilateral carotid occlusion in anesthetized cats. Can J Physiol Pharmacol 50:244–247
29. Lautt WW (1982) Carotid sinus baroreceptor effects on cat livers in control and hemorrhaged states. Can J Physiol Pharmacol 60:1592–1602
30. Maass-Moreno R, Rothe CF (1991) Carotid baroreceptor control of liver and spleen volume in cats. Am J Physiol 260 (Heart Circ Physiol 29):H254–H259
31. Kumada M, Nogami K, Sagawa K (1975) Modulation of carotid sinus baroreceptor reflex by sciatic nerve stimulation. Am J Physiol 228:1535–1541
32. Bell L, Zaret BL, Rutlen DL (1991) Influence of alpha-adrenergic receptor stimulation on splanchnic intravascular volumes in conscious humans. Acta Physiol Scand 143:65–69
33. Nakata K, Leong GF, Brauer RW (1960) Direct measurement of blood pressures in minute vessels of the liver. Am J Physiol 199:1181–1188
34. Shibayama Y, Nakata K (1985) Localization of increased hepatic vascular resistance in liver cirrhosis. Hepatology 5:643–648
35. Lautt WW, Greenway CV, Legare DJ, Weisman H (1986) Localization of intrahepatic portal vascular resistance. Am J Physiol 251:G375–G381
36. Legare DJ, Lautt WW (1987) Hepatic venous resistance site in the dog: localization and validation of intrahepatic pressure measurements. Can J Physiol Pharmacol 65:352–359
37. Lautt WW, Greenway CV, Legare DJ (1987) Effect of hepatic nerves, norepinephrine, angiotensin, and elevated central venous pressure on postsinusoidal resistance sites and intrahepatic pressures in cats. Microvasc Res 33:50–61
38. Bohlen HG, Maass-Moreno R, Rothe CF (1991) Hepatic venular pressures of rats, dogs and rabbits. Am J Physiol 261 (Gastrointest Liver Physiol 24):G539–G547
39. Maass-Moreno R, Rothe CF (1992) Contribution of the large hepatic veins to postsinusoidal vascular resistance. Am J Physiol 262 (Heart Circ Physiol 25):G14–G22
40. Elias H, Popper H (1955) Venous distribution in livers. AMA Arch Pathol 59: 332–340
41. Tucci S Jr, Roquete de Macedo A, Hossne WS (1977) Surgical anatomy of hepatic veins:Morphologic and angiographic study in the dog. Brazilian J Med Biol Res 10:199–204
42. Ungvary G (1977) Functional morphology of the hepatic vascular system. Akademiai Kiado, Budapest

The Effect of Nitroglycerin on the Capacitance of the Human Pulmonary "Venous" System

Senri Hirakawa, Koshi Gotoh, Yukio Ohsumi, Yasuo Yagi, Tatsuo Tsukamoto, Hisato Takatsu, and Yasushi·Terashima[1]

Abstract. We performed a computer simulation of four bouts of successive left-sided heart failure aggravated at intervals of one week. Each bout resulted in a decrease in the level of the left ventricular output curve to 70% of the preceding output curve level, while the right ventricular output curve was maintained unchanged. At each bout of acute left-sided heart failure, there occurred a transient oliguria and a gradual increase in the left atrial pressure. When the ordinate was the pulmonary "venous" volume and the abscissa was left atrial pressure, the pulmonary "venous" volume and left atrial pressure increased linearly (obliquely to the right), with progress of left-sided heart failure every week. By definition, the unstressed volume of the pulmonary "venous" system was unchanged.

It was shown that the occurrence of a simultaneous pulmonary venodilation and systemic venodilation produced a shift, near horizontally to the left, of the volume–pressure relationship.

We reviewed our previous paper where we showed that the volume–pressure (V–P) relation of the human pulmonary "venous" system could be constructed, in the form of V–P lines, by performing radionuclide angiocardiography with passive leg elevation, and using a floating catheter to record the pulmonary capillary wedge pressure (PCW). In 50 patients, most of whom had ischemic heart disease, the sublingual administration of nitroglycerin in a dose of 0.3–0.6 mg caused V–P lines to shift a little downward to the left and become steeper, indicating, in the light of the abovementioned computer simulation, a simultaneous pulmonary and systemic venodilation. The effective unstressed volume of the pulmonary "venous" system was virtually unchanged by nitroglycerin.

Key words: Pulmonary vascular capacitance—Pulmonary "venous" system—Nitroglycerin—Compliance—Unstressed volume

---

[1] Second Department of Medicine, Gifu University School of Medicine, 40 Tsukasa-machi, Gifu, 500 Japan

Introduction

There are few studies on the effect of nitroglycerin on the capacitance of the pulmonary "venous" system [1,2]. Smiseth et al. found that nitroglycerin caused virtually no change in the volume–pressure (V–P) relationship of pulmonary blood vessels, when volume was the pulmonary blood volume and pressure was the microvascular pressure in anesthetized dogs [1]. As to the ·relationship between the estimated pulmonary "venous" volume and mean pulmonary capillary wedge pressure (PCW) in fifty nonanesthetized human subjects, most of whom had ischemic heart disease, we found that the sublingual administration of nitroglycerin in a dose of 0.3–0.6 mg produced a shift and rotation of the V–P relationship which could be interpreted to represent a simultaneous pulmonary and systemic venodilation [2].

The purpose of this article is to review some of the abovementioned findings partly in the light of our previous computer simulation of the pulmonary and/or systemic venodilation superimposed on a chronic left-sided heart failure [3].

Computer Simulation of Acute and Chronic Left-Sided Heart Failure

Figure 1 shows the result of a computer simulation of four bouts of successive left-sided heart failure aggravated at intervals of one week. This was produced by decreasing the left ventricular output curve down to 70% of the preceding output curve level, while maintaining the right ventricular output curve unchanged. The "patient" drank 1 l of water everyday. The digital model used in this simulation was the central, purely cardiovascular part of the famous Guyton model [4], only incorporating the urine volume–systemic arterial pressure relationship of the kidneys that were capable of pressure-diuresis. Also incorporated was our instantaneous resistance-multiplier system capable of increasing the systemic vascular resistance when left ventricular output was decreased [3].

Left ventricular output curve deterioration was associated with oliguria and an increase in the left atrial pressure. The systemic venous compliance (Csv) was purposefully increased by 60% for 3 h on the sixth day every week [3]. The initial systemic venous compliance (CNsv) was 81 ml/mmHg [4], and the unstressed volume of the systemic veins (UVsv), which was held constant throughout the simulation time, was 2970 ml [4]. The response of the left atrial (LA) pressure to the increase in the systemic venous compliance for 3 h on the sixth day was minimal. i.e., LA pressure decreased by about 5 mmHg [3].

Figure 2 also shows the same four bouts of successive left-sided heart failure aggravated at intevals of one week. This time, pulmonary "venous" compliance was increased by 60% for 3 h on the sixth day every week. The initial pulmonary "venous" compliance was 10 ml/mmHg [4] and the unstressed volume of the pulmonary "venous" system (UV p"v"), which was held constant at all

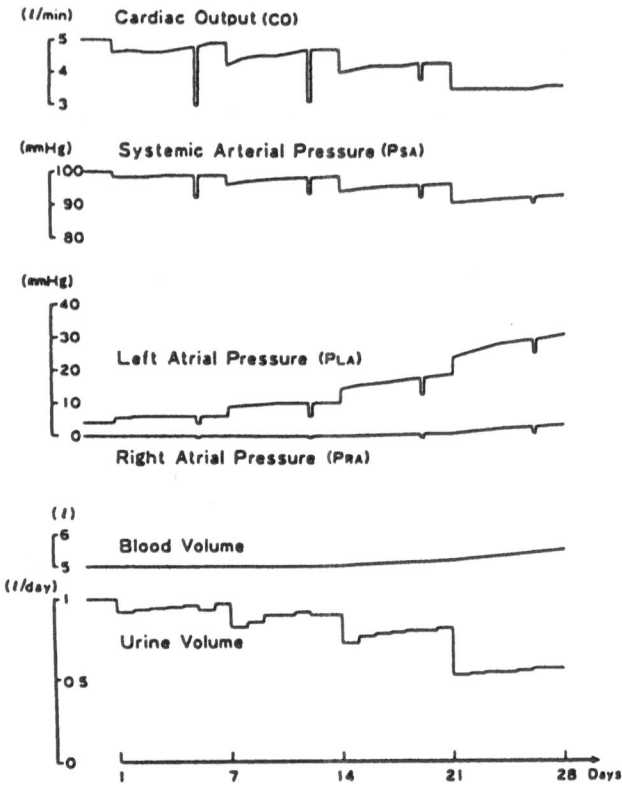

FIG. 1. Simulation of four bouts of successive left-sided heart failure at intervals of one week. Systemic venous compliance was increased by 60% for 3 h on the sixth day every week

times, was 360 ml [4]. The effect of a transient increase in pulmonary "venous" compliance for 3 h on the sixth day was even less striking i.e., LA pressure fell by about 3–4 mmHg [3].

Figure 3 shows the simulated volume–pressure relationship and its alteration with changes in the compliance [3]. The ordinate is the pulmonary "venous" volume and the abscissa is the left atrial pressure. Of the three sets of arrows shown, the smallest arrows show the direction and magnitude of the volume–pressure change that occurs on the sixth day of the first bout of left-sided heart failure. The largest arrows show the direction and magnitude of volume–pressure changes that occur on the sixth day of the fourth bout of left-sided heart failure. The arrows of intermediate size are for the second bout of left-sided heart failure.

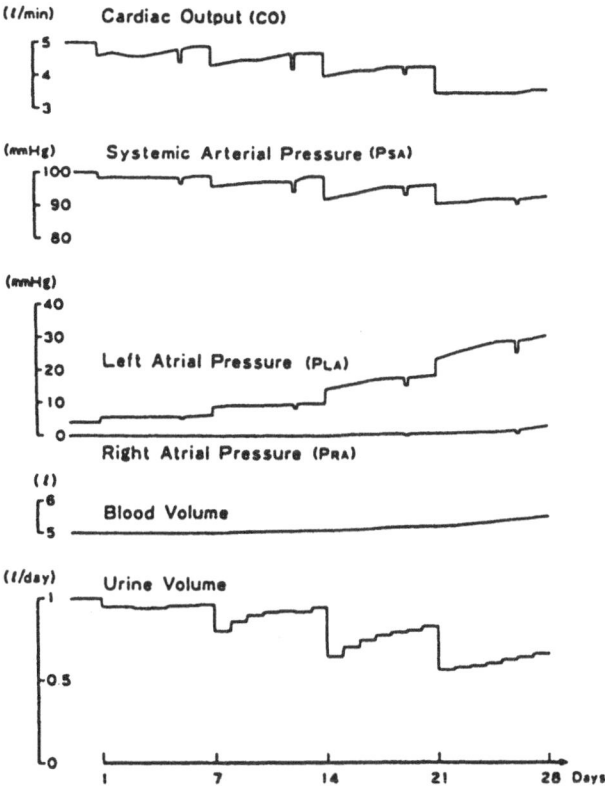

FIG. 2. The same simulation as in Fig. 1. Pulmonary "venous" compliance was increased by 60% for 3 h on the sixth day every week

The Effect of a Combined Pulmonary and Systemic Venodilation on the V–P Relationship of the Pulmonary "Venous" System

By aggravation of left-sided heart failure every week, the pulmonary "venous" volume and the left atrial pressure increased linearly oblique to the right, maintaining the unstressed volume unchanged (360 ml). Figure 3 shows the effect of increasing the systemic venous compliance 1.9 times for 3 h on the sixth day every week ($Csv = CNSV \times 1.9$) and the result was a shift of the P–V plot downward to the left towards the direction of the unstressed volume (360 ml) [3]. Figure 3 also shows the effect of increasing the pulmonary "venous" compliance 1.6 times for 3 h on the sixth day every week ($CP'V' = CNP'V' \times 1.6$) and the result was a shift of the V–P plot upward to the left [3]. The same figure also shows the effect of increasing the pulmonary "venous"

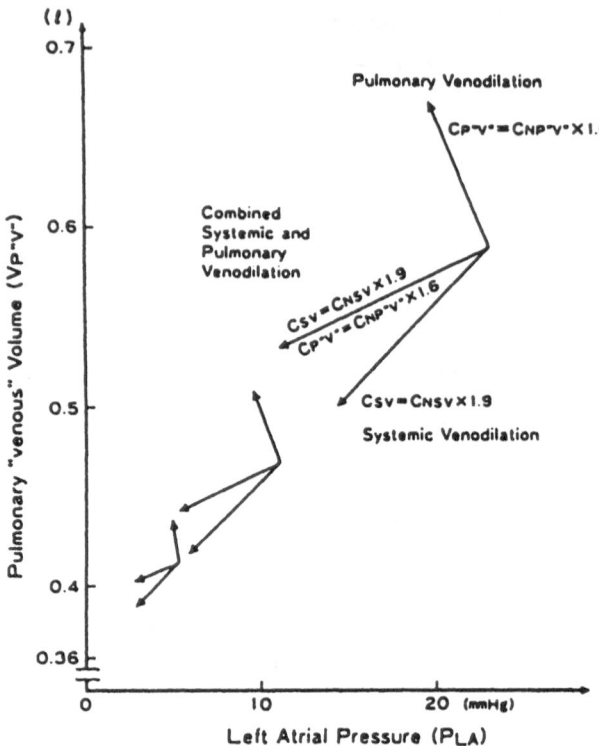

FIG. 3. The same simulation as in Fig.1. Relationship between pulmonary "venous" volume and left atrial pressure is shown. Csv, systemic venous compliance; Cnsv, normal systemic venous compliance; Cp' v', pulmonary "venous" compliance; Cnp' v', normal pulmonary "venous" compliance

compliance 1.6 times, simultaneous with increasing the systemic venous compliance 1.9 times (Cp' v' = Cnp'v' × 1.6 and Csv = Cnsv × 1.9) and the result was a shift to the left, a little downward, of the volume–pressure relationship [3]. It is obvious from this figure that a shift, near horizontally to the left, of the volume–pressure relationship, results from simultaneous pulmonary and systemic venodilation.

The Volume–pressure Relationship in the Pulmonary "Venous" System: Our Previous Study

In a previous paper [2] we showed that the volume–pressure relationship of the human pulmonary "venous" system could be constructed by radionuclide angiocardiography and passive elevation of legs in the form of volume–pressure (V–P) lines where the ordinate gave the pulmonary "venous" volume (P"V"V) and the abscissa gave the pulmonary capillary wedge pressure (PCW). A V–P line was constructed by connecting two plots. One plot was the P"V"V–PCW relationship in baseline (supine) condition. The P"V"V was estimated by P"V"V = 0.7 × PBV, based on Burton's estimation [5]. Pulmonary blood volume (PBV) was measured by radionuclide angiocardiography with Tc99m-erythrocytes. Pulmonary capillary wedge pressure was recorded by a floating catheter. The other V–P plot was generated by measuring ΔP"V"V and ΔPCW associated with passive leg elevation. ΔP"V"V was estimated from the measured PBV and the percentage increase in the radioactivity over the right anterior chest during the leg elevation, in the form of ΔP"V"V = (1/0.8) · PBV · (R_{LE} − R_0)/R_0 where R_0 was the radioactivity over the right anterior chest before leg elevation and R_{LE} was that during the elevation (LE). The term (1/0.8) was a composite correction factor for (1) radioactivity from the anterior and posterior chest wall origin, (2) absorption of radioactive beam within the lung and chest wall, and (3) the share of the pulmonary "venous" system in the overall increase in PBV due to leg elevation. The ΔPCW was an increase in PCW due to leg elevation. The pulmonary "venous" compliance (CP"v") was given by ΔP"V"V/ΔPCW [2].

Figure 4 shows, on the left, a lead plate with a hole in it, so that the region of interest could be set over the right anterior chest at an area away from the heart and liver [2]. Figure 4 also shows, on the right, a typical scintigram indicating the position of the lead plate and the region of interest.

Figure 5 shows typical records of the chest radioactivity with two rounds of leg elevtion before and after the sublingual administration of nitroglycerin. It shows the chest radioactivity (above) and PCW (below) of a 38-year-old man with angina pectoris. PCW is synonymous with pulmonary artery wedge

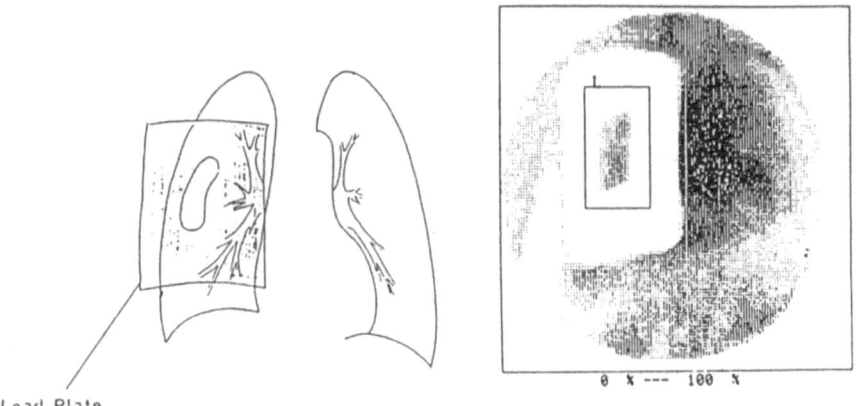

Lead Plate

Fig. 4. After the pulmonary blood volume was determined by radionuclide angiocardiography, the gamma camera was brought over the anterior chest and the lead plate with a hole was set over the right anterior chest, delineating the region of interest

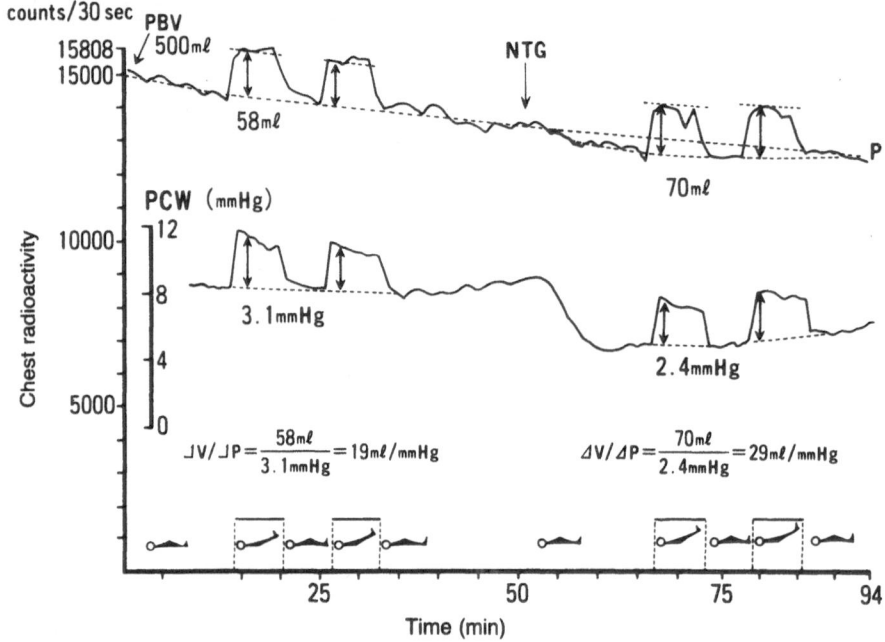

FIG. 5. Typical records of a study on chest radioactivity with leg elevation before and after the sublingual administration of nitroglycerin (*NTG*), showing the chest radioactivity (*above*) and pulmonary capillary wedge pressure (*PCW, below*). *PBV*, pulmonary blood volume; ΔV, change in volume; ΔP, change in pressure

pressure (PAW). The baseline PBV was 500 ml. Legs were elevated twice in succession to show the degree of reproducibility in the change of radioactivity and PCW (PAW) due to leg elevation [2].

Nitroglycerin Produces a Shift and Rotation of the Pulmonary "Venous" System V–P Relationship Compatible with a Simultaneous Pulmonary and Systemic Venodilation

Figure 6 shows a V–P line, AB, before nitroglycerin and another V–P line, CD, after nitroglycerin. A was for the baseline condition and B was recorded during leg elevation. With nitroglycerin, there was a shift of baseline condition from A to C and, with leg elevation, the D plot was generated [2].

Figure 7 shows the P"V"V–PCW plots and their shift with leg elevation, before and after the sublingual administration of nitroglycerin, for 50 patients, most of whom had ischemic heart disease [2]. The compliance of the pulmonary "venous" system averaged 12.4 ± 7.9 (mean ± SD) ml/mmHg before nitroglycerin and 19.2 ± 11.3 ml/mmHg after nitroglycerin ($P < 0.001$). Generally,

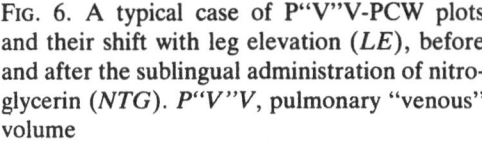

FIG. 6. A typical case of P"V"V-PCW plots and their shift with leg elevation (*LE*), before and after the sublingual administration of nitroglycerin (*NTG*). *P"V"V*, pulmonary "venous" volume

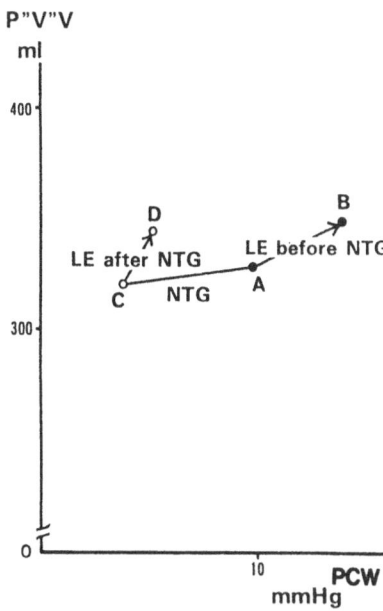

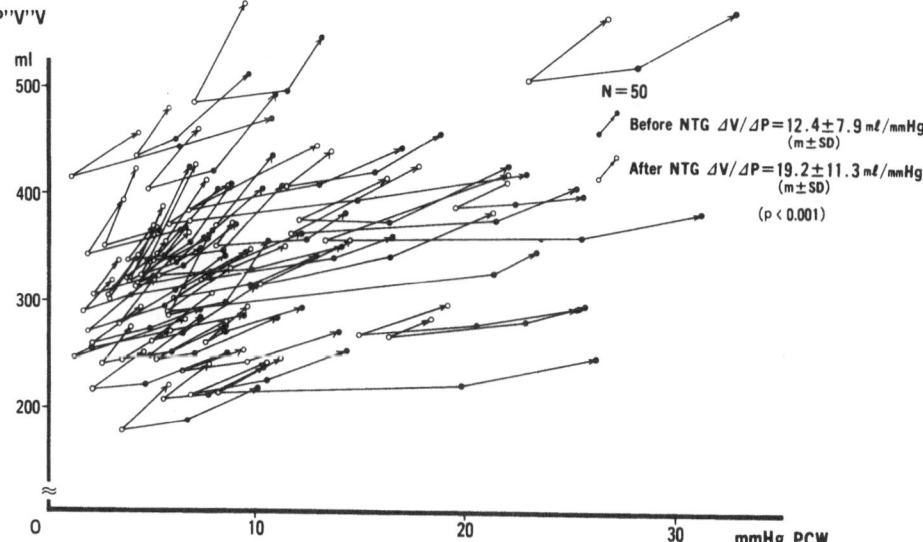

FIG. 7. P"V"V-PCW plots and their shift with leg elevation (*arrow heads*) before (*solid circles*) and after (*open circles*) the sublingual administration of nitroglycerin (*NTG*)

the V–P lines shifted to the left, a little downward, and became steeper after nitroglycerin. This indicated that a simultaneous pulmonary and systemic venodilation took place, in view of the computer simulation of the isolated versus combined pulmonary and systemic venodilation (see Fig. 3).

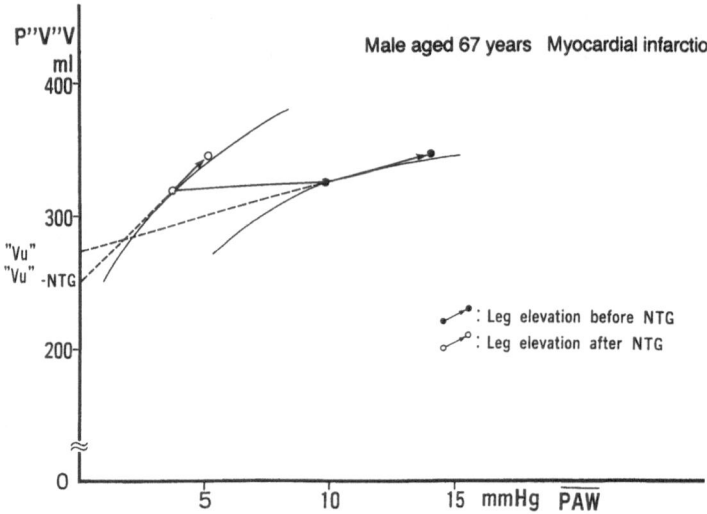

Fig. 8. The pair of solid circles forms a V–P line before nitroglycerin (*NTG*), and its extrapolation determines the effective unstressed volume (Vu) before nitroglycerin. The pair of open circles forms another V–P line after nitroglycerin, and its extrapolation determines the effective unstressed volume after nitroglycerin (*Vu-NTG*). $\overline{PAW}$, pulmonary artery wedge pressure

Figure 8 illustrates the case of a 67-year-old man with myocardial infarction. The pair of solid circles depicts the V–P line before nitroglycerin; extrapolation of this line to zero PCW gave Vu (effective unstressed volume before nitroglycerin). The pair of open circles indicates the V–P line obtained after nitroglycerin; extrapolation to zero PCW gave Vu-NTG (effective unstressed volume after nitroglycerin). The two effective unstressed volumes were not much different [2].

Figure 9 shows the effective unstressed volume (Vu·eff) of the human pulmonary "venous" system before (abscissa) and after (ordinate) the administration of nitroglycerin. Most plots fell within the range of ± 18% coefficient of variation, indicating that effective unstressed volume of the pulmonary "venous" system was virtually unchanged by nitroglycerin [2].

Atrial Natriuretic Peptide (ANP) as a Selective Pulmonary Venodilator

Given in a dose of 0.1 µg/kg per min intravenously to another series of patients, ANP caused the V–P relationship of the human pulmonary "venous" system to shift upward to the left and become steep (increased compliance), indicating a pulmonary venodilation [6], in view of the computer simulation of pure

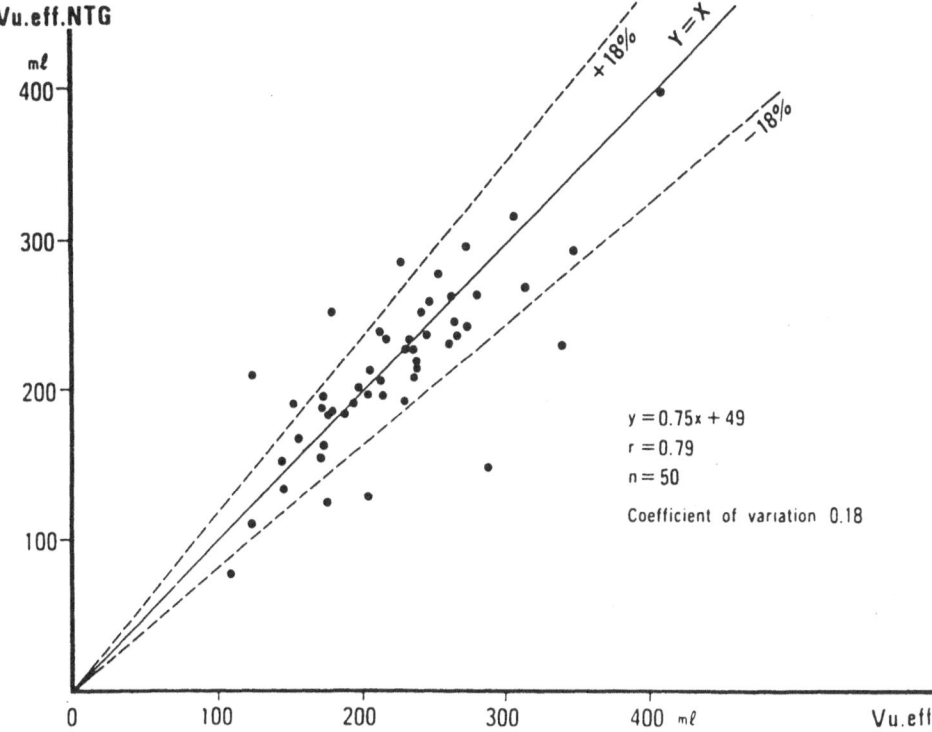

FIG. 9. Effective unstressed volume in human pulmonary "venous" system before (*Vu·eff*) and after (*Vu·eff·NTG*) the sublingual administration of nitroglycerin (*NTG*)

pulmonary venodilation (see Fig. 3). Our preliminary attempts to produce a systemic venodilation as well as this pulmonary venodilation, by giving larger doses of ANP, have not been successful so far, because such dose-elevations induced a diuresis and a consequent decrease in the pulmonary blood volume.

In this sense, ANP appears to be a selective pulmonary venodilator, while nitroglycerin at a typical sublingual dose appears to be a combined pulmonary and systemic venodilator. The fact that nitroglycerin is capable of pulmonary venodilation, basides being effective as a systemic venodilator, is not adequately appreciated. The effect of nitroglycerin on the pulmonary "venous" system has not been described even in a large recent review [7]. Similarities and differences between ANP and nitroglycerin will be the subject for abundant studies when cardiologists begin to use our proposed method of studying the volume–pressure relationships and capacitance of the human pulmonary "venous" system [2].

References

1. Smiseth OA, Manyari DE, Scott-Douglas NW, Wang Y, Kingma I, Smith ER, Tyberg JV (1991) The effect of nitroglycerin on pulmonary vascular capacitance in dogs. Am Heart J 121:1454–1459
2. Hirakawa S, Ohsumi Y, Gotoh K, Suzuki T, Fujiwara H, Yagi Y, Takatsu H (1986) Volume–pressure relations of the human pulmonary "venous" system studied by radionuclide angiocardiography and passive leg elevation, with special reference to the effect of nitroglycerin. Jpn Circ J 50:303–314
3. Kumagai T, Hirakawa S (1983) Computer simulation of congestive heart failure— relaxing effect of nitroglycerin on the human pulmonary "venous" system (in Japanese). Acta Sch Med Universit Gifu 31:225–242
4. Guyton AC, Coleman TG, Granger HJ (1972) Circulation: overall regulation. Ann Rev Physiol 34:13–46
5. Burton AC (1965) Physiology and biophysics of the circulation. Year Book, Chicago, p 64
6. Terashima Y, Gotoh K, Yagi Y, Deguchi F, Nagashima K, Sawa T, Nawada M, Tanaka H, Yasuda N, Hirakawa S (1993) Effects of human atrial natriuretic peptide on the human pulmonary "venous" compliance (abstract in Japanese). Jpn Circ J 57[Suppl I]:441
7. Abrams J (1992) Use of nitrates in ischemic heart disease. Curr Probl Cardiol 17:484–542

Mental Arithmetic Produces Pulmonary and Systemic Venoconstriction

SENRI HIRAKAWA, KOSHI GOTOH, YASUO YAGI, TATSUO TSUKAMOTO, YASUSHI TERASHIMA, KOJI ISHIMURA, and SHINYA MINATOGUCHI[1]

Abstract. We have shown previously that human pulmonary "venous" system volume–pressure relationships (V, ordinate; P, abscissa) can be expressed as a function of the pulmonary "venous" volume, representing V, and the mean pulmonary capillary wedge pressure, representing P. We also described a method for comparing V–P lines before and after the administration of nitroglycerin. In the present report we describe the effect of the performance of mental arithmetic on human pulmonary "venous" system capacitance. By definition, the pulmonary "venous" system consists of pulmonary veins and the left atrium. When subjects were performing mental arithmetic, a nodding response was used because any verbal response resulted in an increase in the pulmonary capillary wedge pressure. Mental arithmetic caused a right-ward shift in the V–P line and a decrease in slope, when compared with controls, indicating pulmonary venoconstriction. In 37 patients, most of whom had ischemic heart disease, the effective unstressed volume increased ($P < 0.01$) and the compliance ($\Delta V / \Delta P$) decreased ($P < 0.01$) during the performance of mental arithmetic. The mean pulmonary capillary wedge pressure in the supine position increased ($P < 0.01$). This decreased compliance indicates pulmonary venoconstriction, while the effective unstressed volume increase suggests an increase in the pulmonary blood volume. Epinephrine and norepinephrine levels in plasma increased during mental arithmetic, suggesting that sympathetic neurohormonal influences were, at least in part, responsible for the observed pulmonary and systemic venoconstriction. We conclude that the human pulmonary "venous" system is under neurohormonal control.

Key words: Compliance—Capacitance—Pulmonary "venous" system—Mental arithmetic

---

[1] Second Department of Medicine, Gifu University School of Medicine, 40 Tsukasa-machi, Gifu, 500 Japan

The Effect of Mental Arithmetic on Cardiovascular Parameters

In determining the cardiovascular responses to the performance of mental arithmetic, recent reports have examined systolic and diastolic blood pressure, heart rate, and other circulatory parameters, including indices of myocardial contractility and cardiac output [1], pre-ejection period and total peripheral resistance [2,3], hemodynamic parameters [4], ascending aortic flow velocity (determined by Doppler echocardiography) [5], pulmonary wedge pressure and stroke volume [6], impedance-derived measurements of cardiac pre-ejection period, stroke index [7], oxygen consumption [8], forearm blood flow [9], cardiac output [10], M-mode echocardiography-derived ejection fraction, left ventricular end-systolic pressure–volume ratio, cardiac output [11], forearm muscle vascular resistance, digital vascular resistance [12], left ventricular filling pressure determined using a Swan-Ganz catheter [13], and forearm vascular resistance [14,15]. We categorized these investigations as class A.

Recent investigations examining the performance of mental arithmetic for its effect on urinary catecholamine excretion have been labeled class B studies [16,17].

Other studies which examined the effect of performance of mental arithmetic on plasma catecholamine levels were categorized as class C studies [8,11,18–22].

Studies in which the cardiovascular effect of the performance of mental arithmetic was examined by determining circulatory parameters other than heart rate and systolic and diastolic blood pressure, and in which blood catecholamine levels were also measured, were classified as class D studies [8,11].

In the present investigation, which was similar to the class D studies, we examined the response of the pulmonary "venous" system capacitance (compliance and the effective unstressed volume) to the performance of mental arithmetic, and we also determined norephinephrine and ephinephrine plasma concentrations.

Human Pulmonary "Venous" System Capacitance and Plasma Norepinephrine and Epinephrine Levels

A total of 37 patients, most of whom had ischemic heart disease, were studied. We have previously reported the influence of drugs, such as nitroglycerin, on the construction of V–P lines, which relate volume (pulmonary "venous" volume, P"V"V) to pressure (mean pulmonary capillary wedge pressure, PCW) [23]. This method employed Tc99m-erythrocyte radionuclide angiocardiography, a floating catheter to record the PCW, and passive leg elevation. We recently compared the latter method to an entirely different method, using phasic PCW measurements in order to determine pulmonary "venous" system compliance ($\Delta V/\Delta P$).

A composite V–P line was generated from two V–P plots. The initial V–P graph plot was created by measuring the pulmonary blood volume (PBV) using radionuclide angiocardiography and Stewart-Hamilton's principle (based partly on Burton's data [24]), such that P"V"V = 0.7 × PBV, and by determining the PCW using a floating catheter. The second V–P plot was generated by measuring the ΔP"V"V and ΔPCW during passive leg elevation, where ΔP"V"V was estimated as ΔP"V"V = 0.8 × ΔPBV, and ΔPCW was the increase in PCW during passive leg elevation. The compliance (CP"v") was defined as the slope of the V–P line (ΔV/ΔP). This was closely correlated with compliance determined using a method which mathematically analyzed the PCW wave-forms ($n = 41$, unpublished observations). In the P"V"V–PCW plane where a P"V"V ordinate and a PCW abscissa were plotted, the V–P line was ex-trapolated to zero PCW, and the effective unstressed volume was defined as the V-intercept [23]. The location and slope of the V–P line can be expressed by three parameters, i.e., compliance, effective unstressed volume, and PCW.

Antecubital vein blood samples, obtained before, during, and after mental arithmetic, were examined for plasma norepinephrine (NE) and epinephrine (E) levels by high performance liquid chromatography (HPLC) using the THI method as published elsewhere [25].

The Effect of Mental Arithmetic on Pulmonary "Venous" System Capacitance

In a typical case, a V–P plot (Fig. 1) was generated from a 59-year-old man with an old myocardial infarction, at rest (C) and after leg elevation (LE). After returning to a supine position, mental arithmetic was monitored using a nodding response. The performance of mental arithmetic resulted in a shift of the V–P plot (designated MA), and leg elevation (LE) in addition to mental arithmetic resulted in a V–P plot shift, designated MA+LE. It appears that mental arithmetic causes a V–P line shift upwards and to the right, resulting in a slight clockwise rotation.

Figure 2 represents the average V–P line, in the control and mental arith-metic periods (C to LE, and MA to MA+LE), respectively. Thus, mental arithmetic is associated with a right-ward and upward V–P line shift and a clockwise rotation. We therefore conclude that mental arithmetic is associated with pulmonary venoconstriction. This is a novel finding.

Table 1 shows various pulmonary "venous" system capacitance parameters before and during mental arithmetic.

The compliance (CP"v") was 14.1 ± 6.4 ml/mmHg (mean ± SD) during the control period and decreased to 8.7 ± 3.8 ml/mmHg during mental arithmetic ($P < 0.01$). The effective unstressed volume (Vu·eff) increased during mental arithmetic ($P < 0.01$).

The PCW in the supine position also increased from 8.3 ± 4.1 mmHg during the control period to 11.2 ± 5.2 mmHg during mental arithmetic ($P < 0.01$).

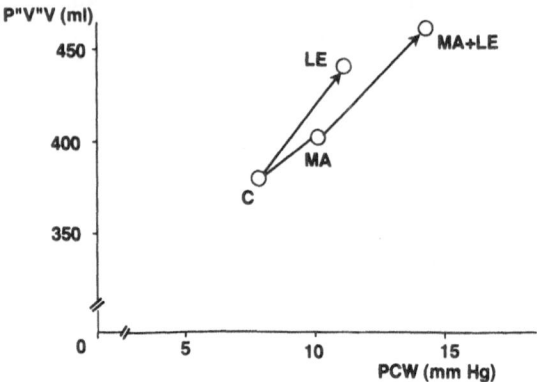

FIG. 1. Volume–pressure (V–P) lines in a 59-year-old man with an old myocardial infarction. The control V–P plot (*C*) was obtained in the supine position. Passive leg elevation also generated a V–P plot (*LE*). After the patient was restored to the supine position, mental arithmetic commenced, monitored using a nodding response, generating a V–P plot (*MA*). Leg elevation in addition to mental arithmetic generated the V–P plot designated *MA+LE*. It appears that, with mental arithmetic, there was a rightward and upward V–P line shift, giving a slight clockwise rotation. PCW, pulmonary capillary wedge pressure; P"V"V, pulmonary "venous" volume

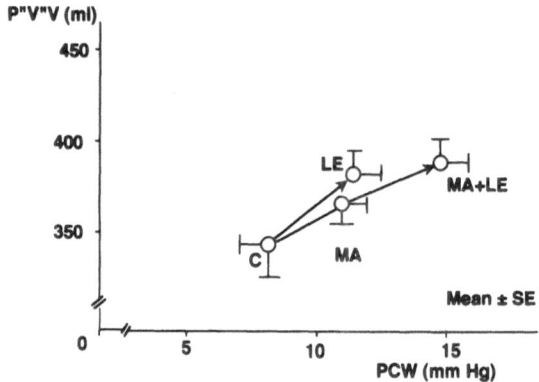

FIG. 2. Mean V–P lines (*n* = 27), during the C to LE, and during the C to MA, and the MA to MA+LE period. Using the data from Table 1, the small increase in the PCW between C and MA is statistically significant (*P* < 0.01), but the small increase in P"V"V between C and MA is not statistically significant

The observed decrease in compliance indicates that mental arithmetic produces pulmonary venoconstriction. Moreover, the observed increase in Vu·eff suggests that mental arithmetic results in a volume shift from the systemic to the pulmonary vascular bed.

TABLE 1. Changes in various pulmonary "venous" system capacitance parameters with mental arithmetic.

	Unit	Before mental arithmetic	During mental arithmetic	P value
Compliance	ml/mmHg	14.1 ± 6.4*	8.7 ± 3.8	P < 0.01
Vu·eff	ml	237 ± 64	271 ± 72	P < 0.01
PCW	mmHg	8.3 ± 4.1	11.2 ± 5.2	P < 0.01
P"V"V	ml	348 ± 75	365 ± 84	NS
NE (n = 25)	µg/ml	0.300 ± 0.135	0.382 ± 0.200	P < 0.01
E (n = 26)	µg/ml	0.060 ± 0.051	0.120 ± 0.092	P < 0.01

Vu·eff, effective unstressed volume; *PCW*, pulmonary capillary wedge pressure; *P"V"V*, pulmonary "venous" volume; *NE*, norepinephrine; *E*, epinephrine
* Numbers refer to the mean ± SD

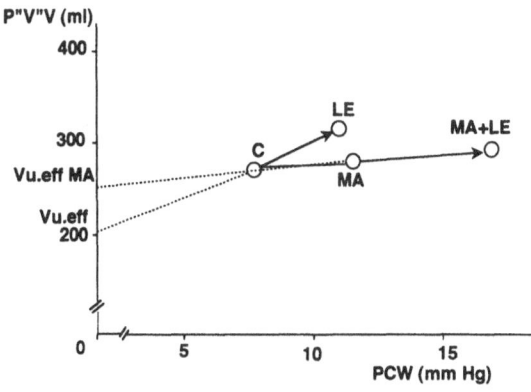

FIG. 3. The V–P line shift and rotation, due to mental arithmetic, in 65-year-old man with an old myocardial infarction. This figure specifically shows that the effective unstressed volume (Vu·eff) was greater after mental arithmetic than before it

Figure 3 represents a 65-year-old man with an old myocardial infarction, in which the mental arithmetic resulted in an increased pulmonary "venous" system Vu·eff.

The Effect of Mental Arithmetic on Plasma Norepinephrine (NE) and Epinephrine (E) Levels

The plasma NE and E levels were altered in most of the study patients (Table 1). Both NE and E levels increased significantly during mental arithmetic, but fully returned to their pre-mental arithmetic levels after the completion of mental arithmetic (data not shown). This result indicates that mental arithmetic is associated with sympathetic stimulation. We suggest that this sympathetic neurohormonal influence is, at least partly, responsible for the observed pulmonary, and suggested systemic, venoconstriction.

References

1. Sherwood A, Davis MR, Dolan CA, Light KC (1992) Cardiovascular reactivity assessment: effects of choice of difficulty on laboratory task responses. Int J Psychophysiol 12:87–94
2. Turner JR, Sherwood A, Light KC (1991) Generalization of cardiovascular response: supportive evidence for the reactivity hypothesis. Int J Psychophysiol 11:207–212
3. Allen MT, Boquet AJ Jr, Shelley KS (1991) Cluster analysis of cardiovascular responsivity to three laboratory stressors. Psychosom Med 53:272–288
4. Fasano ML, Gasta I, De Simon G, Iannuzzi R, Ferrara LA (1991) Cardiovascular response to adrenergic stimulation during treatment with tertatolol. A new noncardioselective beta-blocking agent in primary hypertensive patients. Jpn Heart J 32:435–444
5. Pannarale G, Isea JE, Coats AJ, Conwey J, Sleight P (1991) Cardiac and blood pressure responses to mental stress in reactive hypertensives. Clin Exp Hypertens A 13:1–12
6. Mazzuero G, Temporelli PL, Tavaggi L (1991) Influence of mental stress on ventricular pump function in postinfarction patients. An invasive hemodynamic investigation. Circulation 83[Suppl IV] II:145–154
7. Kasprowicz AL, Manuck SB, Malkoff SB, Krantz DS (1990) Individual differences in behaviorally evoked cardiovascular response: temporal stability and hemodynamic patterning. Psychophysiology 27:605–619
8. Blumenthal JA, Fredrikson M, Kuhn CM, Ulmer RL, Walsh Riddle M, Appelbaum M (1990) Aerobic exercise reduces levels of cardiovascular and sympathoadrenal responses to mental stress in subjects without prior evidence of myocardial ischemia. Am J Cardiol 65:93–98
9. Anderson NB, Lane JD, Taguchi F, Williams RB Jr (1989) Patterns of cardiovascular responses to stress as a function of race and parental hypertension in men. Health Psychol 8:525–540
10. Allen MT, Crowell MD (1989) Patterns of autonomic response during laboratory stressors. Psychophysiology 26:603–614
11. Kaji Y, Ariyoshi K, Tsuda Y, Kanaya S, Fujino T, Kuwabara H (1989) Quantitative correlation between cardiovascular and plasma epinephrine response to mental stress. Eur J Appl Physiol 59:221–226
12. Goldstein HS, Edelberg R, Meier CF, Davis L (1989) Relationship of expressed anger to forearm muscle vascular resistance. J Psychosom Res 33:497–504
13. Mazzuero G, Zotti AM, Bertolotti G, Tavazzi L (1989) Hemodynamic response to different types of mental stress in patients with recent myocardial infarction. Jpn Heart J 30:35–46
14. Anderson NB, Lane JD, Taguchi F, Williams RB Jr, Houseworth SJ (1989) Race, parental history of hypertension, and patterns of cardiovascular reactivity in women. Psychophysiology 26:39–47
15. Santangelo K, Falkner B, Kushner H (1989) Forearm hemodynamics at rest and stress in borderline hypertensive adolescents. Am J Hypertens 2:52–56
16. Fredrickson M, Tuomisto M, Bergman Losman B (1991) Neuroendocrine and cardiovascular stress reactivity in middle-aged normotensive adults with parental history of cardiovascular disease. Psychophysiology 28:656–664

17. Aldo Ferrara L, Soro S, Mainenti G, Mancini M, Pisanti N, Borrelli R, Moscato T, Mancini M (1989) Body weight and cardiovascular response to sympathetic stimulation in childhood. Int J Obes 13:271–277

18. Suarez EC, Williams RB Jr, Kuhn CM, Zimmerman EH, Schanberg SM (1991) Biobehavioral basis of coronary-prone behavior in middle-aged men. Part II: Serum cholesterol, the Type A behavior pattern, and hostility as interactive modulators of physiological reactivity. Psychosom Med 53:528–537

19. Mills PJ, Dimsdale JE, Ziegler MG, Berry CC, Bain RD (1990) Beta-adrenergic receptors predict heart rate reactivity to a psychosocial stressor. Psychosom Med 52:621–623

20. Pomerlean CS, Pomerlean OF, McPhee K, Morrell EM (1990) Discordance of physiological and biochemical response to smoking and to psychological stress. Br J Addict 85:1309–1316

21. Mills PJ, Schneider RH, Dimsdale JE (1989) Anger assessment and reactivity to stress. J Psychosom Res 33:379–382

22. Lenders JW, Willemsen JJ, de Boo T, Lemmens WA, Thien T (1989) Disparate effects of mental stress on plasma noradrenaline in young normotensive and hypertensive subjects. J Hypertens 7:317–323

23. Hirakawa S, Ohsumi Y, Gotoh K, Suzuki T, Fujiwara H, Yagi Y, Takatsu H (1986) Volume–pressure relations of the human pulmonary "venous" system studied by radionuclide angiocardiography and passive leg elevation, with special reference to the effect of nitroglycerin. Jpn Circ J 50:303–314

24. Burton AC (1965) Physiology and biophysics of the circulation. Year Book, Chicago, p 64

25. Ishimura K, Ito H, Minatoguchi S, Suzuki S, Watanabe H, Miyauchi Y, Hirakawa S (1988) Response of peripheral venous pressure and plasma catecholamine concentration to supine leg exercise—A study in patients with mild congestive heart failure. Jpn Circ J 52:119–130

Human Pulmonary Venous Return Curve: Effect of Dopamine

Kohshi Gotoh, Yasuo Yagi, Hisato Takatsu, Yasushi Terashima, Kenshi Nagashima, Toshiyuki Sawa, Fumiko Deguchi, Masumi Nawada, Haruhito Tanaka, Hiroyasu Ito, and Senri Hirakawa[1]

Abstract. Pulmonary "venous" compliance ($C_P"v"$) and pulmonary arterial compliance (C_{PA}) were determined using radionuclide angiocardiography and a floating catheter in patients with various cardiac diseases. A pulmonary venous return curve was constructed from these measurements, and the effect of dopamine on the pulmonary venous return curve was evaluated. At the same time the pulmonary "venous" volume–pressure relationship and its response to dopamine was documented.

Dopamine increased $C_P"v"$, indicating the dilation of the pulmonary "venous" system. Dopamine also increased the mean pulmonary pressure (P_{MP}) and shifted the pulmonary venous return curve parallel to the right, probably owing to a shift of blood from the systemic to the pulmonary vascular bed.

Key words: Human pulmonary venous return curve—Pulmonary "venous" compliance—Pulmonary vascular capacitance—Dopamine

Introduction

The pulmonary venous system plays an important role as a reservoir for the left ventricle. According to Guyton et al. [1], cardiac output is determined by an equilibrium between the cardiac output curve and the venous return curve. Similarly, an equilibrium point exists between the left ventricular output curve and the pulmonary venous return curve. The latter has an almost straight downslope region that crosses the left atrial pressure axis at the point of mean pulmonary pressure (P_{MP}). Therefore, a pulmonary venous return curve can roughly be drawn by connecting the point of P_{MP} on the left atrial pressure axis and the point of intersection of the left atrial pressure and the cardiac output (Fig. 1). It is necessary to estimate P_{MP} for the purpose of drawing a pul-

[1] Second Department of Internal Medicine, Gifu University School of Medicine, Tsukasa-machi 40, Gifu 500, Japan

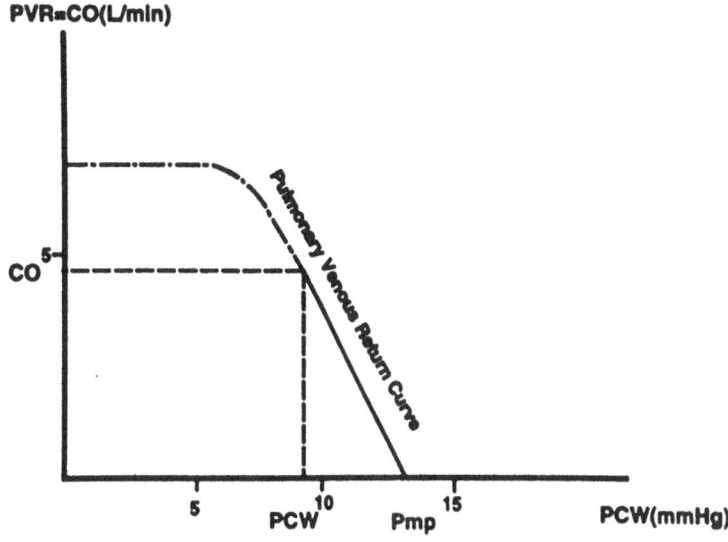

FIG. 1. Pulmonary venous return (*PVR*) curve. *Solid line*, pulmonary venous return curve drawn by connecting the point of mean pulmonary pressure (PMP) and the equilibrium point. *CO*, cardiac output; *PCW*, pulmonary capillary wedge pressure

monary venous return curve. According to Guyton et al., PMP is defined as follows [1]:

$$P_{MP} = (EV_{P"v"} + EV_{PA})/(C_{P"v"} + C_{PA}) \qquad (1)$$

where $EV_{P"v"}$ and EV_{PA} express the extra volumes (stressed volume) in the pulmonary "venous" system and pulmonary arterial system, respectively, whereas $C_{P"v"}$ and C_{PA} represent the compliances of the pulmonary "venous" system and pulmonary arterial system, respectively. We constructed the down-sloping part of the pulmonary venous return curve by connecting the point of PMP on the abscissa and the point of CO-PCW plot. The left atrial pressure was measured as the pulmonary capillary wedge pressure (PCW). This article aims to describe the effect of dopamine on the human pulmonary "venous" volume–pressure relationships and that on the human pulmonary venous return curve. By definition, the pulmonary "venous" system consists of the pulmonary veins and the left atrium.

Measurement of Pulmonary Blood Volume and Measurement of Pulmonary "Venous" Compliance

Ten patients, nine men and one woman, aged 52.2 ± 10.1 (mean ± SD), with various cardiac diseases, consented to undergo the examination. The final diagnoses were old myocardial infarction (4 patients), angina pectoris (3 patients), and cardiomyopathy (3 patients). Three of them were in class I and the rest were in class II of the cardiac functional classification for dyspnea of the previous New York Heart Association (NYHA) classification.

Patients were placed on the scintigraphic table in the supine position at rest, and radionuclide angiocardiography was performed using a gamma camera (ZLC, Siemens) equipped with a parallel hole collimator in the left anterior oblique view, and injection of 720 MBq of Tc99m-pertechnetate to label in vivo the red blood cells with stannous pyrophosphate. Data were processed with a dedicated computing system (Scintipac 2400, Shimazu, Japan). Pulmonary blood volume (PBV) was determined by Eq. 2, using the indicator dilution method:

$$PBV = CO \cdot \Delta MTT_{PAB\text{-}LA} \qquad (2)$$

where CO is cardiac output, and $\Delta MTT_{PAB\text{-}LA}$ represents mean transit time of the injected radioactive tracer from the pulmonary artery bifurcation to the left atrium. In some cases in whom left atrial radioactivity was difficult to separate from that of the other chambers on angiocardiographic images, we used the peak-to-peak time (PPT) of the time–activity curve recordable from the region of interest covering the entire cardiac silhouette. Concerning the PPT in this large area in relation to the $\Delta MTT_{PAB\text{-}LA}$, we had already found the following equations [2] (Fig. 2):

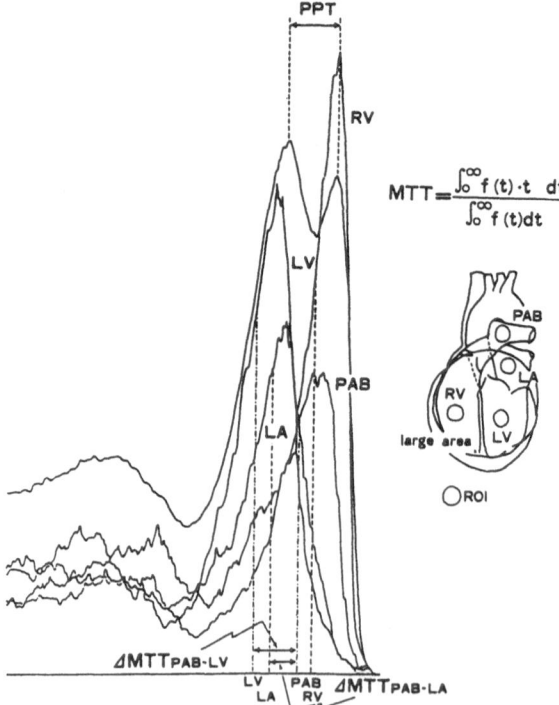

$$MTT = \frac{\int_0^\infty f(t) \cdot t \; dt}{\int_0^\infty f(t) dt}$$

FIG. 2. Time–activity curve obtained with a gamma camera by setting the region of interest window (*ROI*) over various regions of the cardio-pulmonary system: left ventricle (*LV*); right ventricle (*RV*); pulmonary artery bifurcation (*PAB*); left atrium (*LA*). $\Delta MTT_{PAB\text{-}LV}$ represents the mean transit time from the pulmonary artery bifurcation to the left ventricle. *PPT*, peak-to-peak time

$$\Delta MTT_{PAB-LV} = PPT_{large\ area} \tag{3}$$

$$\Delta MTT_{PAB-LA} = \Delta MTT_{PAB-LV} \cdot 0.77 + 0.05 \tag{4}$$

$$\therefore \Delta MTT_{PAB-LA} = PPT_{large\ area} \cdot 0.77 \tag{5}$$

Therefore, we may write

$$PBV = CO \cdot PPT_{large\ area} \cdot 0.77 \tag{6}$$

By definition, pulmonary venous volume (P"V"V) including the left atrium was 70% of PBV [3]:

$$P"V"V = 0.7 \cdot PBV \tag{7}$$

After equilibrium of the tracer was achieved, a region of interest was delineated by a hole in a lead plate placed on the patient's right middle chest away from the heart and liver, and the radioactivity in this region of interest was recorded continuously. The mean pulmonary capillary wedge pressure (PCW), considered to represent the internal pressure in the pulmonary "venous" system, was also recorded simultaneously with a floating catheter. Thus, in the supine position, a volume (P"V"V)–pressure (PCW) plot was generated. After the radioactivity from the region of interest was ascertained to be stable in the supine position, both legs of the patients were passively elevated to 30° in order to increase the blood volume in the pulmonary vascular system. The time-

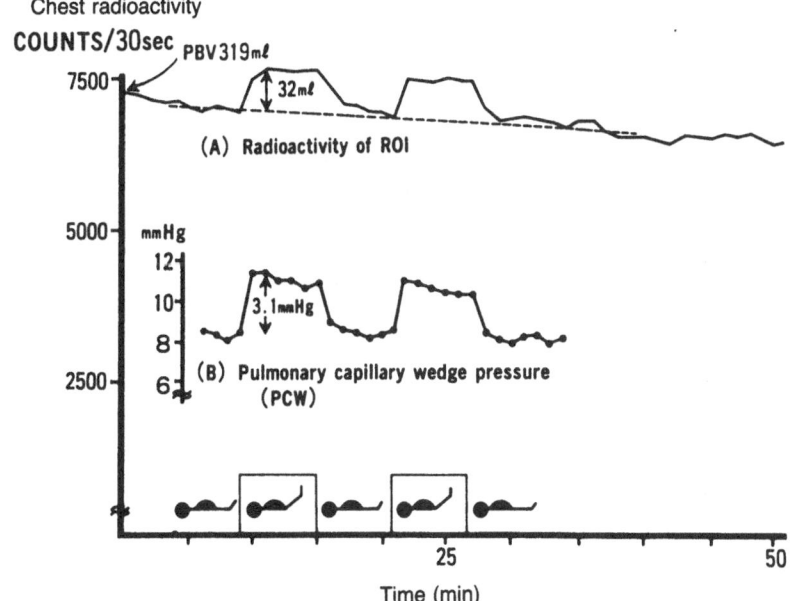

Fig. 3. Typical time-course of change of radioactivity in the region of interest, the right middle lung (*ROI, upper-trace*), and PCW (*lower-trace*), at rest and with leg elevation. *PBV*, pulmonary blood volume

course of radioactivity and of PCW in one case is shown in Fig. 3. We assumed that the increment in the pulmonary "venous" volume (ΔP"V"V) was 80% of the increment of pulmonary blood volume (ΔPBV):

$$\Delta P\text{"V"}V = 0.8 \cdot \Delta PBV \tag{8}$$

We have shown in a previous report [4] that ΔP"V"V during leg elevation could be calculated using Eq. 9:

$$\Delta P\text{"V"}V = (1/0.8) \cdot PBV \cdot (R_{LE}\text{-}R_0)/R_0 \tag{9}$$

(R_{LE}, radioactivity during leg elevation; R_0, radioactivity in supine position.) Eq. 9 was defined after correcting for chest wall activity and the attenuation caused by the lung and chest wall [4].

Pulmonary "venous" compliance (CP"v") was calculated by dividing the increment of pulmonary "venous" volume during leg elevation (ΔP"V"V) by the increment of the pulmonary capillary wedge pressure (ΔPCW), i.e., CP"v" = ΔP"V"V/ΔPCW) [4]. These tests were performed before and during the intravenous administration of dopamine at rate of 5 μg/kg per min.

Measurement of Pulmonary Arterial Compliance and the Calculation of PMP

Pulmonary arterial compliance (CPA) was obtained from the diastolic decline of pulmonary arterial pressure using Levenson's equation [5]:

$$CPA = DT/\{R \cdot \ln(R_s/R_d)\} \tag{10}$$

where R is total pulmonary resistance, and P_s and P_d are end-systolic and end-diastolic pressure in the pulmonary artery, respectively.

According to Guyton et al. [1], PMP can be calculated from Eq. 1. The extra volumes in the pulmonary arterial system (EVPA) and in the pulmonary "venous" system (EVP"v") were calculated using the values of CPA and CP"v", measured as just described, and mean pulmonary artery pressure (P_{PA}):

$$EVPA = CPA \cdot P_{PA} \tag{11}$$

$$EVP\text{"v"} = CP\text{"v"} \cdot PCW \tag{12}$$

The Effect of Dopamine on the Pulmonary "Venous" System

Table 1 shows the effect of dopamine on the circulatory parameters and the pulmonary vascular capacitance.

When one plots the pulmonary "venous" volume (P"V"V) on the ordinate and PCW on the abscissa before and during the passive leg elevation, the line connecting these two points is a V–P line, i.e., a short segment of volume–

TABLE 1. Parameters of pulmonary "venous" system and hemodynamic data.

	Control	Dopamine
Heart rate (beats/min)	69.9 ± 10.3	74.1 ± 13.0*
Cardiac index (l·min^{-1}·m^2)	2.75 ± 0.46	3.27 ± 0.66**
Pulmonary capillary wedge pressure (PCW)(mmHg)	9.6 ± 2.5	11.3 ± 3.5
Pulmonary artery compliance (CPA)(ml/mmHg)	3.62 ± 1.07	3.72 ± 1.37
Pulmonary "venous" compliance (CP"v")(ml/mmHg)	9.8 ± 4.8	15.1 ± 7.1**
Pulmonary "venous" volume (P"V"V)(ml)	309.6 ± 67.6	330.7 ± 72.7*
Mean pulmonary pressure (PMP)(mmHg)	11.6 ± 2.4	13.4 ± 3.9**

$n = 10$, mean ± SD, **$P < 0.01$, *$P < 0.05$

FIG. 4. Short segment of volume–pressure curve in the pulmonary venous system before and during dopamine infusion in a 64-year-old male patient with angina pectoris (5 µg/kg per min). Dopamine shifted the curve up and to the right, and increased the slope. *P"V"V*, pulmonary "venous" volume

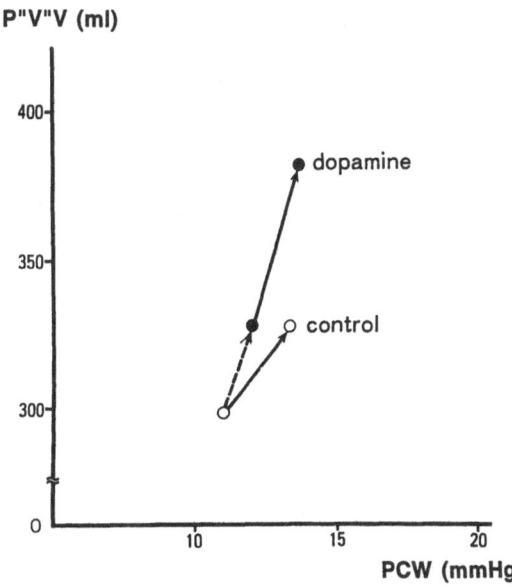

pressure curve in the pulmonary "venous" system. Figure 4 shows V–P lines obtained in one case. Dopamine shifted the pulmonary "venous" V–P line up and to the right, and the slope became steeper (increased compliance). The average pulmonary "venous" compliance (CP"v") significantly increased, from 9.8 ± 4.8 ml/mmHg to 15.1 ± 7.1 ml/mmHg ($P < 0.01$). This suggests that dopamine shifted blood from the systemic into the pulmonary vascular bed, while it caused pulmonary venodilation.

As to the pulmonary venous return (PVR) curve, in the PVR(CO)–PCW plane, the point of PMP on the abscissa and the point of CO-PCW plot were connected by a straight line. We were thus able to draw the down-sloping part of the pulmonary venous return curve. Figure 5 represents the pulmonary venous return curve before and during dopamine administration. With dopamine, the curve was shifted parallel to the right.

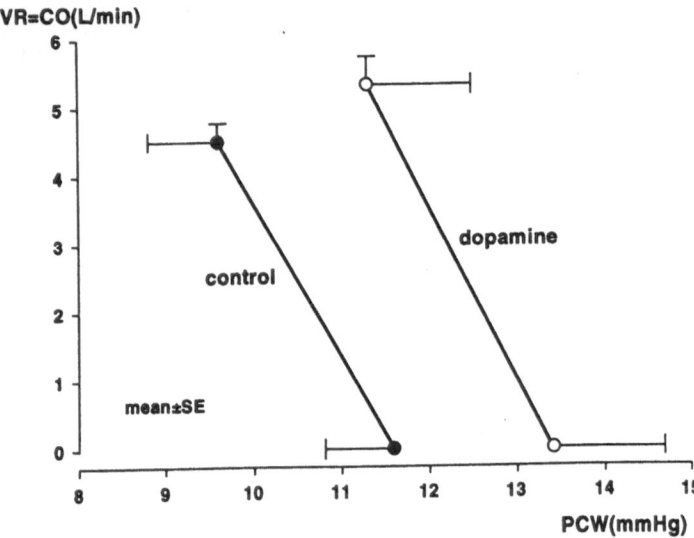

Fɪɢ. 5. Down-sloping parts of pulmonary venous return (*PVR*) curves before and during dopamine infusion. Dopamine causes a rightward shift of the curve, while the slope remains almost parallel to the control. *CO*, cardiac output

Implications of the Study

Other than our preliminary report [6], there has been no report of success in constructing the pulmonary venous return curve in the human, nor even in animal studies. Usually, the change in cardiac output is discussed in terms of a cardiac output curve. However, it is important to discuss it in terms of both the left ventricular output curve and the pulmonary venous return curve. In other words, the concept of the pulmonary venous return curve becomes important when one evaluates the influence of vascular tonus and blood volume change, on the cardiac output. In the present study, dopamine shifted the human pulmonary venous return curve parallel to the right (Fig. 5), probably due to the increase in blood volume in the pulmonary circulatory system, in spite of the fact that dopamine caused a pulmonary venodilation (Fig. 4).

References

1. Guyton AC, Jones CE, Coleman TC (1973) Mean circulatory pressure, mean systemic pressure, and mean pulmonary pressure and their effect on venous return. In: Circulatory physiology: Cardiac output and its regulation. Saunders, Philadelphia, pp 205–221

2. Goto K, Hirano A, Hirakawa S (1981) Non-invasive estimation of the human pulmonary blood volume with gamma camera and RI-angiography. Jpn Circ J 45:113–119
3. Burton AC (1972) Arrangements of many veins. In: Physiology and biophysics of the circulation, 2nd edn Year Book. Medical Publisher, Chicago, pp 51–62
4. Hirakawa S, Ohsumi Y, Gotoh K, Suzuki T, Fujiwara H, Yagi Y, Takatsu H (1986) Volume–pressure relations of the human pulmonary "venous" system studied by radionuclide angiocardiography and passive leg elevation, with special reference to the effect of nitroglycerin. Jpn Circ J 50:303–314
5. Levenson JA, Safer ME, Simon AC, Kheder AI, Daou JN, Levy BI (1981) Systemic arterial compliance and diastolic run-off in essential hypertension. Angiology 32:402
6. Terashima Y, Gotoh K, Yagi Y, Takatsu H, Deguchi F, Nagashima K, Sawa T, Tanaka H, Nawada M, Hirakawa S (1993) Pulmonary venous return curves in left-sided heart disease (in Japanese). Heart 25[Suppl I]:90–96

Pulmonary Vein Flow Velocity-Time Profile for Semiquantitative Estimates of Left Atrial Storage Fraction

Michio Arakawa, Hiroshi Miwa, Yoshimi Ito, Kensaku Kagawa, Toshiyuki Noda, Kazuhiko Nishigaki, Masaaki Tomita, Ryuhei Tanaka, and Senri Hirakawa[1]

Abstract. To evaluate the usefulness of a method of estimating the ratio of left atrial (LA) storage volume to left ventricular (LV) stroke volume (i.e., the LA storage fraction) by Doppler echocardiography, we compared the LA storage fraction from the pulmonary vein (PV) flow velocity–time profile by Doppler echocardiography with the estimate obtained by conventional cineangiocardiography in 19 patients. From cineangiocardiograms, we calculated the LA storage fraction as the ratio of the LA storage volume to LV stroke volume. From the flow velocity–time profile of the left upper PV recorded by transesophageal Doppler echocardiography, we measured the flow velocity–time integrals during systole and diastole (Sa, Da). Provided that the cross-sectional area remains relatively unchanged during one cardiac cycle, the flow velocity–time integral is equivalent to the volumetric flow rate–time integral. The LA storage fraction was calculated as Sa/(Sa+Da). The LA storage fraction from cineangiocardiography was 0.57 ± 0.10 (mean $\pm$ SD), and 0.64 ± 0.08 from Doppler echocardiography, and the difference was not significant. Although estimates of the LA storage fraction from Doppler echocardiography tended to be slightly higher than those from cineangiocardiography, the PV flow velocity–time profile obtained by Doppler echocardiography appears to be clinically useful.

Key words: Pulmonary vein—Flow velocity-time integral—Left atrium—Storage fraction—Doppler echocardiography—Cincangiocardiography

The Concept and Significance of Left Atrial Storage Fraction

During ventricular systole, the right ventricular stroke volume is stored in a lumped vascular bed ending with the closed mitral valve and comprising the pulmonary arterial system, the capillaries, the pulmonary venous (PV) system,

---

[1] The Second Department of Internal Medicine, Gifu University School of Medicine, 40 Tsukasa-machi, Gifu, 500 Japan

and the left atrium (LA). The volume stored in the LA during ventricular systole is defined as the LA storage volume, all of which in turn is conveyed to the left ventricle (LV) during ventricular diastole. The LA storage volume assists LV filling through storage of an appropriate amount of blood and generation of an appropriate driving pressure for LV early filling (i.e., LA passive release), and subsequently through maintenance of an appropriate preload (an appropriate volume and pressure) for the LA active contraction for LV late filling (i.e., LA active release). The significance of the LA storage volume can be evaluated in terms of a normalized variable. One of the normalizations is the ratio of LA storage volume to right ventricular stroke volume, which is defined as the LA storage fraction.

Methods of Estimating Left Atrial Storage Fraction

The "gold-standard" method of estimating the LA storage fraction is cineangiocardiography for the LA and LV [1–3]. In a steady state, the right ventricular stroke volume is equal to the LV stroke volume. Accordingly, the LA storage fraction is determined as the ratio of LA storage volume to LV stroke volume.

However, since cineangiocardiography is considerably invasive and complex, a less invasive and simpler method would be desirable. Using transesophageal Doppler echocardiography, we can easily and reliably record the left upper PV flow velocity–time profile. The PV flow velocity–time profile would be equal to the PV volumetric flow rate–time profile under two major assumptions: (1) the cross-sectional area of the PV remains unchanged during systole and diastole; (2) the PV flow velocity–time profile is uniform among four major PVs opening to the LA. If so, any one of the PV flow velocity–time profiles would be representative for the LA volumetric flow rate–time profile. We can obtain the LA storage fraction as the ratio of the PV flow velocity–time integral during ventricular systole (Sa) to the sum of the PV flow velocity–time integrals during systole and diastole (Sa plus Da), namcly Sa/(Sa + Da).

To validate the usefulness of this new method of estimating the LA storage fraction from the PV flow velocity–time profile, we compared the figures from Doppler echocardiography with those obtained by cineangiocardiography in the same patient.

A Comparison of Methods of Estimating Left Atrial Storage Fraction: Cineangiocardiography and Doppler Echocardiography

The diagnoses of the 19 patients studied were a history of arrhythmia ($n = 7$), cardiomyopathy ($n = 3$), coronary artery disease ($n = 2$), and others ($n = 7$) (Table 1). Patients with mitral regurgitation or aortic regurgitation of greater-

TABLE 1. Patient profile, and hemodynamic, cineangiocardiographic, and echocardiographic variables.

Patient number	Diagnosis	Age (years)	mPAW (mmHg)	LAG LASV (ml)	LAG & LVG LASV/SV	TEE Sa/(Sa+Da)
1	VPC	52	4	43	0.65	0.64
2	SSS	22	8	53	0.50	0.57
3	VPC	18	9	37	0.46	0.65
4	DCM	54	6	40	0.52	0.53
5	CAD	69	7	49	0.62	0.64
6	Aortitis	34	9	45	0.65	0.66
7	VPC	61	7	33	0.59	0.58
8	Pericarditis	58	5	40	0.56	0.58
9	HCM	67	7	43	0.52	0.66
10	HCM	30	7	32	0.41	0.50
11	VPC	59	8	28	0.39	0.46
12	HTN	66	9	78	0.65	0.70
13	CAD	62	6	49	0.58	0.65
14	HTN	59	5	67	0.66	0.64
15	Sarcoidosis	68	6	37	0.82	0.79
16	Chest pain	44	10	49	0.50	0.68
17	Chest pain	67	10	30	0.58	0.76
18	SSS	61	10	51	0.63	0.61
19	VPC	53	13	62	0.62	0.65
	Mean	52.8	7.7	45.6	0.57	0.64
	± SD	15.8	2.2	12.9	0.10	0.08

LAG, left atriography; LVG, left ventriculography; TEE, transesophageal echocardiography; mPAW, mean pulmonary artery wedge pressure; LASV, left atrial storage volume; SV, left ventricular stroke volume; LASV/SV and Sa/(Sa + Da), left atrial storage fraction of ventricular stroke volume from cineangiography and Doppler echocardiography, respectively; Sa, velocity–time integral of S wave; Da, velocity–time integral of D wave; CAD, coronary artery disease; DCM, dilated cardiomyopathy; HCM, hypertrophic cardiomyopathy; HTN, hypertension; SSS, sick sinus syndrome; VPC, ventricular premature contraction

than-mild severity by Doppler color flow signal were excluded. Most patients had cardiac catheterization first, and then transesophageal Doppler echocardiography within a week.

Single-plane left ventriculography and biplane LA angiocardiography were performed. Cineangiocardiograms were filmed at 50 frames/s. The LV volume was calculated using the method of Kennedy et al. [4]. The LA volume devoid of the LA appendage [5] was calculated by the Simpson's rule method using a film motion analyzer (nac CARDIAS 300, nac Inc.). The LA volume was then corrected with the regression equation derived from our study using 10 human cadaver casts of the LA (true volume ranged from 9 to 163 ml): true volume (ml) = 0.91 × calculated volume−1.1 ($r = 0.995$, $P < 0.01$) [6]. The LV stroke volume was calculated as the LV maximum volume minus the LV minimum volume, while the LA storage volume as the LA maximum volume minus the LA minimum volume at the end of LA active contraction. Finally, the LA storage fraction was calculated as the ratio of LA storage volume to LV stroke volume.

By transesophageal Doppler echocardiography (HP77020AC, 5-MHz transducer, Hewlett Packard Co.), the PV flow velocity–time profile was recorded. The sample volume (1.2 mm) was placed at the center of the bloodstream in the left upper PV slightly distal to the orifice into the LA, where the Doppler color flow signal showed the maximum flow velocity. The Doppler beam was aligned as parallel as possible to the long axis of the blood flow and any angle correction was within 30°. The PV flow velocity–time integrals during ventricular systole (Sa) and diastole (Da) were measured using a planimeter. The onset of diastolic flow was defined as the nadir between the systolic and diastolic waves. Three consecutive cardiac cycles were analyzed and averaged. The LA storage fraction was calculated as the ratio of the PV flow velocity–time integral during systole (Sa) to the sum of the PV flow velocity–time integrals during systole and diastole (Sa + Da), namely Sa/(Sa + Da).

The paired Student's t-test was used and $P < 0.05$ was accepted as significant. The linear least-squares method was used for a regression equation. Limits of agreement were examined by the method of Bland and Altman [7].

The representative PV flow velocity–time profile is shown in Fig. 1. Patient profiles, and hemodynamic, Doppler echocardiographic, and cineangiocardiographic data are shown in Table 1. The LA storage fraction was 0.57 ± 0.10 (mean ± SD) by cineangiocardiography and 0.64 ± 0.08 by Doppler echocardiography. The difference between the two estimates was not significant.

Figure 2 is a scatter plot which shows that the LA storage fraction measured by Doppler echocardiography (y) tended to be larger than the LA storage fraction from cineangiocardiography (x), yielding the regression equation: $y = 0.569x + 0.302$ ($r = 0.710$, $P < 0.01$). The correlation coefficient is influenced by the range of values studied. Thus, to examine the degree of

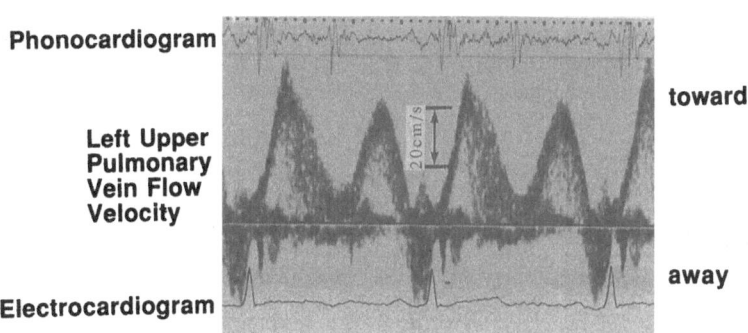

Fig. 1. Pulmonary vein flow velocity–time profile. The patient is a 58-year-old male

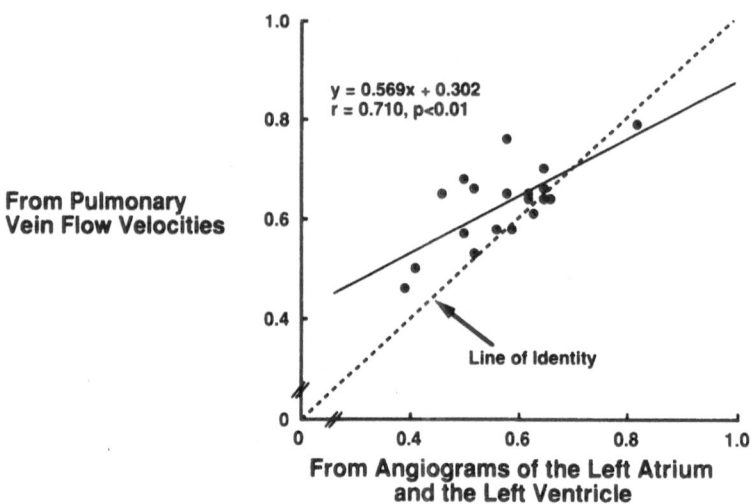

FIG. 2. A scatter plot of the left atrial storage fraction derived from the pulmonary vein flow velocity–time profile (*y* axis) against that from the cineangiocardiogram (*x* axis)

agreement of values between the cineangiocardiography and the Doppler echo-cardiography, we compared the difference between the two values with the average of these two values (i.e., the best estimate of the true values). The mean difference was 0.05, but the limits of agreement (i.e., 2 standard deviations of the difference) were $+0.19$ and -0.09, which indicates that the value from Doppler echocardiography may be 0.19 above or 0.09 below the value obtained by cineangiocardiography. In conclusion, transesophageal Doppler echocardiography appears to be clinically useful for a semiquantitative estimate of the LA storage fraction.

Studies on the LA Storage Fraction by Angiocardiography

By biplane angiocardiography obtained at 6 or 12 exposures per second, the overall average of the LA storage fraction was 0.42 in patients with aortic stenosis or mitral stenosis [8].

By biplane cineangiocardiography, the LA storage fraction was 0.48 ± 0.10 (mean $\pm$ SD) in subjects with a normal heart and 0.65 ± 0.17 in patients with myocardial infarction [1]. The LA storage fraction was calculated to be 0.47 ± 0.12 in patients with coronary artery disease [2]. The LA storage fraction was calculated to be 0.51 ± 0.08 at rest and 0.81 ± 0.10 during balloon occlusion of the coronary artery in patients with coronary artery disease [3]. The LA storage fraction (*y*) is reported to increase with age in subjects with a normal heart, yielding the regression equation: $y = 0.00331 \times \text{age} + 0.309, r = 0.563$ [6]. From this study, it follows that the LA storage fraction is 0.51 at the age of

60. These previous studies indicate that the LA storage fraction measured by the present cineangiocardiography is within a range of reasonable values.

Studies on the LA Storage Fraction by Doppler Echocardiography

From the recording of the PV blood flow velocity–time profile by transesophageal Doppler echocardiography, Kuecherer et al. measured the ratio of the systolic velocity–time integral to the sum of the systolic and diastolic velocity–time integrals, which they termed the systolic fraction [9]. Their PV systolic fraction, which is equal to the LA storage fraction in the present study, correlated negatively and significantly with mean LA pressure, yielding the regression equation: mean LA pressure (mmHg) = 35 − 0.39 × LA storage fraction (%) ($r = -0.88$). It follows that at mean LA pressures of 10 mmHg and 20 mmHg, the LA storage fraction is 0.64 and 0.38, respectively.

The LA storage fraction, which was estimated from the PV flow velocity–time integral obtained by transesophageal Doppler echocardiography, increased with age in noncardiac surgical subjects under general anesthesia, yielding the regression equation: LA storage fraction = 0.00373 × age + 0.514 ($r = 0.656$, $P < 0.01$) [10]. It follows that the LA storage fraction is 0.65 and 0.74 at the age of 40 and 60, respectively. These figures are considerably different from the estimate using cineangiocardiography (0.51 vs 0.74 at the age of 60). However, the LA storage fraction estimated by the present Doppler echocardiography would be comparable to the LA storage fraction from other Doppler echocardiographic studies [9,10].

Incidentally, Basnight et al. [11] recorded the PV flow velocity–time profile by transthoracic Doppler echocardiography, from which we can calculate the mean LA storage fraction to be 0.60.

Some Unsolved Assumptions in the PV Flow Velocity–Time Profile

Phasic Changes in Cross-Sectional Area of the PV

From the animal study by Rajagopalan et al. [12], the cross-sectional area of the extraparenchymal PV can be calculated to change about 30% during one cardiac cycle at a normal LA pressure. At a higher LA pressure, the cross-sectional area of the extraparenchymal PV changes less, and the cross-section changes from a dumbbell shape to a circle. In addition, the change in cross-sectional area occurs in both systole and diastole, and possibly to an equal extent. This would indicate that the velocity–time profile would be affected to an equal extent in both systole and diastole. Accordingly, the time course of the products of instantaneous cross-sectional area and velocity, which reflects

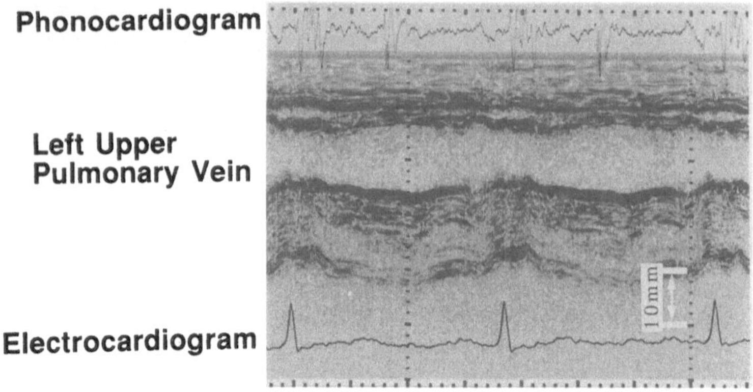

Phonocardiogram

**Left Upper
Pulmonary Vein**

Electrocardiogram

FIG. 3. An M-mode echocardiographic record of the left upper pulmonary vein by transesophageal echocardiography. Assuming that the cross-section of the pulmonary vein is circular, the change in the cross-sectional area is 54% between the maximum and the minimum. The left atrial storage fraction is 0.56 from the pulmonary vein flow velocity–time integral and 0.52 from the pulmonary vein volumetric flow rate–time integral. The patient is a 58-year-old male

the volumetric flow rate–time profile, appears to be equally affected in both systole and diastole. Therefore, we can fairly reasonably assume that the LA storage fraction from the PV flow velocity–time integral equals the LA storage fraction from the PV volumetric flow rate–time integral.

We partly substantiated our afore-mentioned assumptions using the original recording of the M-mode echogram of the left upper PV obtained by chance (Fig. 3). Assuming that the cross-section of the PV is circular and that the distance between anterior and posterior walls of the PV is a diameter, the cross-sectional area decreased to 54% of its maximum cross-sectional area. The volumetric flow rate (product of the instantaneous cross-sectional area and velocity at 40-ms intervals)–time profile was calculated. The LA storage fraction was 0.56 from the flow velocity–time integral and 0.52 from the volumetric flow rate–time integral. Therefore, we conclude that the LA storage fraction tends to be larger based on the PV flow velocity–time integral than on the PV volumetric flow rate–time integral, but the PV flow velocity–time integral provides a good estimate of the LA storage fraction.

Regurgitant Volume Due to LA Active Contraction

In a steady state, LA active contraction regurgitates a certain amount of volume to the PV. This regurgitant volume is conveyed together with real systolic runoff of right ventricular stroke volume to the LA during the LA filling period (i.e., during LV systole). Therefore, the regurgitant volume should be separated from the real LA storage volume. However, this regurgitant volume is unavoidably included in the LA storage volume in the method of cineangiocardiography. In this study, we ignored a component of the velocity–time integral during LA active contraction (Aa) in the estimation of the LA storage fraction from the PV flow velocity–time profile. However, to more exactly match the conditions between the two methods, we calculated the LA storage fraction as Sa/(Sa − Aa + Da) in 19 subjects; this value was larger than that calculated by the present method (Sa/(Sa + Da)) (0.66 ± 0.09 vs 0.64 ± 0.08).

Other Limitations of the Present Study

Both studies were not performed simultaneously. Intervals were up to one week between cineangiocardiography and Doppler echocardiography, which would change the hemodynamic state. In addition, the two examinations seem to impose different levels of stress, which might change the hemodynamic state.

The angiographic measurement of LA volume is not fully validated yet, partly because the volume of the LA appendage is excluded as in this study. In addition, we do not know for certain whether the LA assumes an ideal ellipsoidal shape during one cardiac cycle, particularly at the end of LA active contraction.

The sampling site of the PV flow velocity is subjected to movement with cardiac contraction and respiration. Therefore, the sampling site would deviate, from time to time, from the axial stream of the laminar flow in the PV.

Although the PV flow velocity–time integrals in the right upper PV and in the left upper PV were not significantly different during either the systolic or diastolic wave [13], this does not validate our assumption that the flow velocity pattern is practically equal in all four PVs.

Regardless of these complex limitations, we believe that the PV flow velocity–time profile seems to be useful as a semiquantitative estimate of the LA storage fraction.

References

1. Matsuda Y, Toma Y, Ogawa H, Matsuzaki M, Katayama K, Fujii T, Yoshino F, Moritani K, Kumada T, Kusukawa R (1983) Importance of left atrial function in patients with myocardial infarction. Circulation 67:566–571
2. Yamaguchi M, Arakawa M, Tanaka T, Takaya T, Nagano T, Hirakawa S (1987) Study on left atrial contractile performance – Participation of Frank-Starling mechanism. Jpn Circ J 51:1001–1009
3. Sigwart U, Gerbic M, Goy J-J, Kappenberger L (1990) Left atrial function in acute transient left ventricular ischemia produced during percutaneous transluminal coronary angioplasty of the left anterior descending coronary artery. Am J Cardiol 65:282–286

4. Kennedy JW, Trenholme SE, Kasser JS (1970) Left ventricular volume and mass from single-plane cineangiocardiogram. A comparison of anteroposterior and right anterior oblique methods. Am Heart J 80:343–352

5. Sauter HJ, Dodge HT, Johnston RR, Graham TP (1964) The relationship of left atrial pressure and volume in patients with heart disease. Am Heart J 67:635–642

6. Miwa H, Arakawa M, Kagawa K, Noda T, Nishigaki K, Ito Y, Hirakawa S (1991) Estimation of systolic storage volume ratio of the left atrium with cineangiography. Circulation 84(Suppl II):44

7. Bland JM, Altman DG (1986) Statistical methods for assessing agreement between two methods of clinical measurement. Lancet 1:301–310

8. Grant C, Bunnell IL, Greene DG (1964) The reservoir function of the left atrium during ventricular systole. An angiocardiographic study of atrial stroke volme and work. Am J Med 37:36–43

9. Kuecherer HF, Muhiudeen IA, Kusumoto FM, Lee E, Moulinier LE, Cahalan MK, Schiller NB (1990) Estimation of mean left atrial pressure from trans-esophageal pulsed Doppler echocardiography of pulmonary venous flow. Circulation 82:1127–1139

10. Arakawa M, Akamatsu S, Terazawa E, Dohi S, Miwa H, Kagawa K, Nishigaki K, Ito Y, Hirakawa S (1992) Age-related increase in systolic fraction of pulmonary vein flow velocity–time integral from transesophageal Doppler echocardiography in subjects without cardiac disease. Am J Cardiol 70:1190–1194

11. Basnight MA, Gonzalez MS, Kershenovich SC, Appleton CP (1991) Pulmonary venous flow velocity: relation to hemodynamics, mitral flow velocity and left atrial volume, and ejection fraction. J Am Soc Echocardiogr 4:547–558

12. Rajagopalan B, Bertram CD, Stallard T, Lee G de J (1979) Blood flow in pulmonary veins: III Simultaneous measurements of their dimensions, intravascular pressure and flow. Cardiovasc Res 13:684–692

13. Castello R, Pearson AC, Lenzen P, Labovitz AJ (1991) Evaluation of pulmonary venous flow by transesophageal echocardiography in subjects with a normal heart. Comparison with transthoracic echocardiography. J Am Coll Cardiol 18:65–71

Venous Disorders of the Leg Evaluated by a Plethysmographic Technique

MASAFUMI HIRAI[1]

Abstract. In the present study, the venous function of the leg was evaluated by strain gauge plethysmography, and the quantitative relationships of venous hemodynamics with the aggravation of varicosities was investigated. The expelled volume during five active dorsiflexions of the feet, and the venous recovery time, were calculated to evaluate muscle pump efficiency and valvular competence. Chronic venous insufficiency was characterized by a shortened recovery time and a high value for the expelled volume using a tourniquet. From these results, it was concluded that the main pathogenesis of varicose veins of the leg was a high degree of venous congestion of the calf due to valvular incompetence of the superficial vein system, and the aggravation caused the chronic venous insufficiency. The abnormal increase of venous pooling decreased after beneficial treatments such as a stripping operation or the wearing of compression stockings.

Key words: Varicose veins—Chronic venous insufficiency—Strain gauge plethysmography—Stripping operation—Elastic stockings

Introduction

In our previous studies [1–3], strain gauge plethysmography was used to evaluate venous function of the leg, in order to investigate the pathophysiology of venous disorders and to study the effect of traditional beneficial treatments on venous dysfunction. This paper is a review of our previous presentations [1–3].

---

[1]Department of Surgery, Aichi Prefectural College of Nursing, Tougoku, Uesidami, Moriyama, Nagoya, 463 Japan

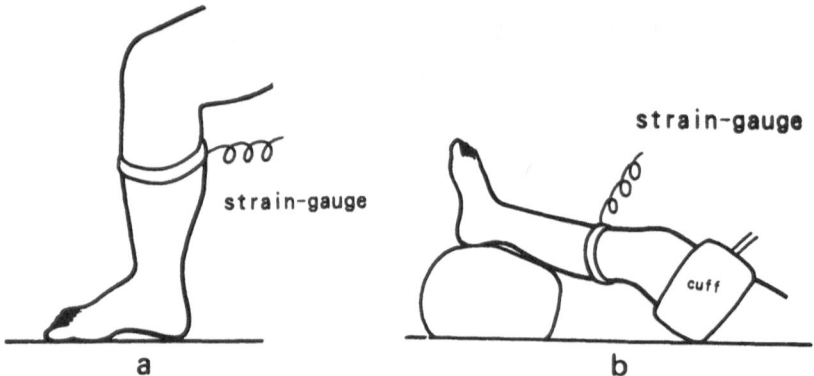

FIG. 1a,b. Application of strain gauge plethysmography. a Muscle pump plethysmography; b venous occlusion plethysmography

Hemodynamic Comparison of Normal Veins, Primary Varicosities, and Deep Vein Thrombosis

In strain gauge plethysmography, two application techniques [4] have been reported for evaluating the venous function of the leg—muscle pump plethysmography and venous occlusion plethysmography (Fig. 1):

1. Muscle Pump Plethysmography

In this method, the subjects were studied in the sitting position on a chair. Both feet were placed on the floor with the lower extremities slightly apart. The strain gauge transducers were applied at the midcalf. The subjects were instructed to exercise their feet five times forcefully. The recording was continued until a stable horizontal baseline was attained. During the exercise, the baseline decreased according to the decrease of blood volume of the calf. After the cessation of the exercise, the baseline increased again. To evaluate the effect of the superficial saphenous system, rubber tourniquets were applied just below the knee. The exercise was then repeated with the tourniquets in place.

The curves were analyzed for: (1) half-refilling time ($\frac{1}{2} \cdot t$), which is defined as the number of seconds necessary for the baseline shift to reach 50% of the steady state after the exercise; and (2) the expelled volume during the exercise (EV, ml/dl), which was calculated from the maximum volume changes during the exercise divided by 1% calibration (Fig. 2).

This study included 32 normal limbs, 104 limbs with primary varicose veins (varicosis group), and 34 limbs with postthrombotic syndrome (thrombosis group). In all limbs with venous disorders, the competence of the deep vein system was evaluated by ascending phlebography and physical examinations.

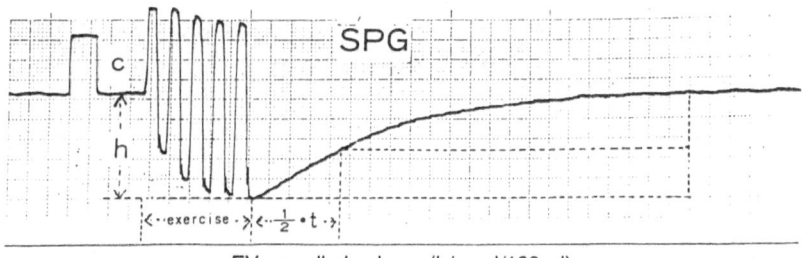

EV: expelled volume (h/c, ml/100 ml)
1/2·t: half refilling time (s)

FIG. 2. Parameters in muscle pump plethysmography

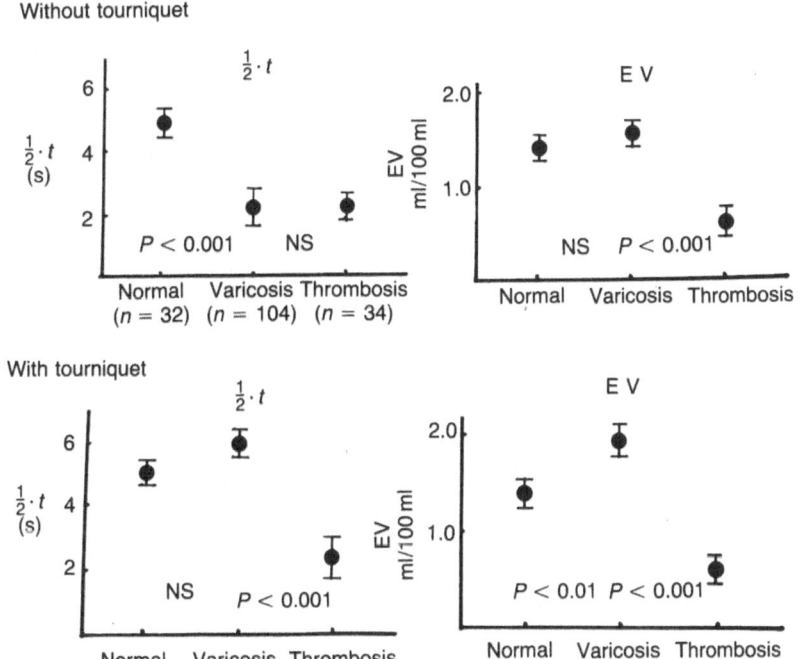

FIG. 3. Results of muscle pump plethysmography (mean ± SE). $\frac{1}{2}$ · t, half-refilling time; EV, expelled volume. The significance of the difference between adjacent groups is indicated

Figure 3 shows the results of muscle pump plethysmography. The value of $\frac{1}{2}$ · t in varicosis and thrombosis groups shortened significantly as compared with the normal group. With an application of tourniquets below the knee, the $\frac{1}{2}$ · t value in the varicosis group increased significantly, and subsequently, the significant difference with the normal group disappeared.

These results indicate that in the varicosis group, the reflux of the superficial vein system plays an important role in the pathogenesis.

The value of EV in the varicosis group was significantly greater than that in the thrombosis group. There was no significant difference between normal and varicosis groups. With an application of a tourniquet, however, the value in the varicosis group significantly increased, and then became greater than that of normal group.

The value of EV might be influenced by several factors, including contraction of calf muscles, arterial inflow, function of venous valves, the competence of the deep vein system, and congestion of blood in the leg.

The high value of EV with tourniquets observed in the varicosis group might result from a large quantity of blood accumulation in the calf, caused by high venous reflux [5]. When the amount of blood pooling in the calf is large, a large volume of blood is expelled towards the heart during calf muscle contractions, i.e., a high EV results.

2. Venous Occlusion Plethysmography

In this technique, the subjects are studied in the supine position, and the strain gauge is applied at the midcalf and the cuff at the thigh. The cuff is inflated to a pressure of about 50 mmHg. The baseline gradually increases and reaches a plateau after 2 or 3 minutes. Then the cuff pressure is suddenly released. From the curve obtained, two parameters—venous capacitance (VC, ml/dl) and maximum venous outflow (MVO, ml/dl per min)—are calculated (Fig. 4) [4]. In this study, the VC, which was calculated from the maximum volume increase during the cuff inflation divided by 1% calibration, was used.

Figure 5 shows the results of venous occlusion plethysmography. The highest value of VC was observed in the varicosis group, as reported by other authors [6]. This result indicates an abnormal increase of distensibility of vein walls in the varicosis group.

From these findings, it is concluded that the pathogenesis of varicose veins might be characterized as a high degree of venous congestion of the leg and valvular incompetence of the superficial vein system.

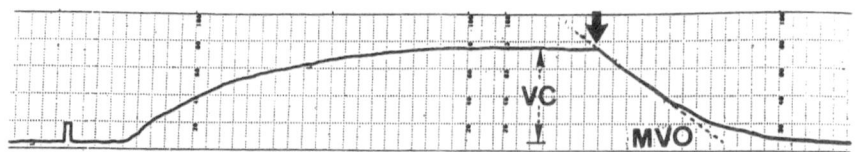

FIG. 4. Parameters in venous occlusion plethysmography. Venous capacitance (VC) is calculated from the maximum volume increase during the cuff inflation divided by 1% calibration. MVO, maximal venous outflow

FIG. 5. Venous capacitance in three different groups (mean ± SE). The significance of the difference between adjacent groups is indicated

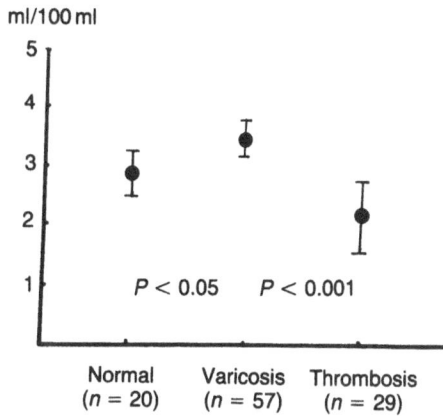

Hemodynamic Changes Before and After Treatments

Stripping Operation

In the present study, venous function was evaluated by strain gauge plethysmography before and six months after the varicose vein surgery. This technique was applied in 46 limbs with varicose veins. As the control, venous function was also evaluated in 40 normal limbs.

Surgery was performed under general anesthesia. The procedure [2] consisted of extraction of the long saphenous vein from the groin to the ankle, ligation of incompetent communicating veins, and avulsion of varicose tributaries via multiple small incisions.

Before the surgery, the $\frac{1}{2} \cdot t$ in the varicosis group was significantly shorter than in normal controls. The EV showed no significant difference between the two groups. After application of a tourniquet, the varicosis group showed an improvement in $\frac{1}{2} \cdot t$, but subsequently the significant difference between the two groups disappeared. The EV also increased significantly with the use of a tourniquet, to become significantly larger than that of the normal control group.

Six months after the operation, the $\frac{1}{2} \cdot t$ improved significantly (Fig. 6), meaning the abnormal reflux of the superficial veins improved after the treatment.

The EV showed no significant changes between measurements before and after the operation. However, when the tourniquet was applied below the knee, no significant increase was observed at postoperative examination. These results indicate that the high expelled volume obtained with a tourniquet disappears after the surgery.

The fact that the EV showed no significant changes with improvement in $\frac{1}{2} \cdot t$ after the surgery might be explained by two opposing effects of the treatment,

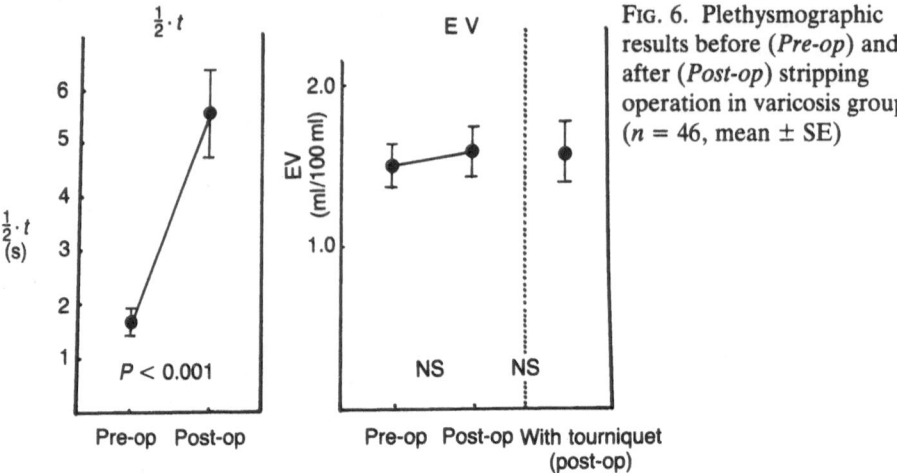

FIG. 6. Plethysmographic results before (*Pre-op*) and after (*Post-op*) stripping operation in varicosis group ($n = 46$, mean ± SE)

a reduction in reflux and a decrease in venous pooling in the lower limbs [3]. The former results in an increase in EV and the latter in a decrease.

Elastic Stocking

Muscle pump plethysmography was used to assess the effect of elastic stockings on 10 normal limbs and 52 limbs with varicose veins [3]. After the routine measurements, as described in the previous section, the subjects were instructed

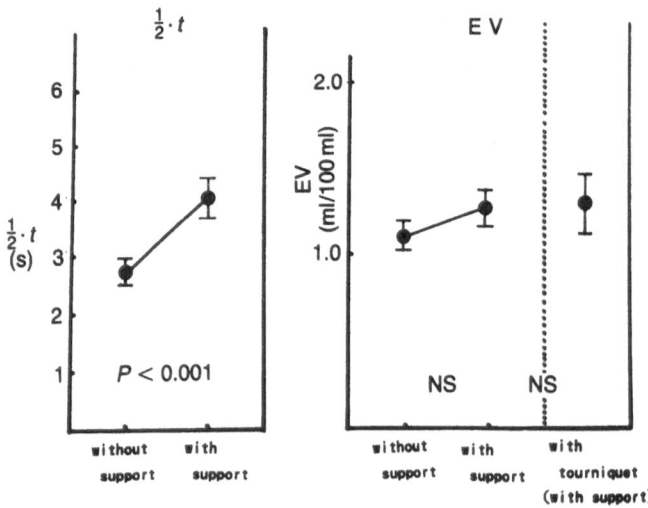

FIG. 7. Plethysmographic results without and with elastic support in varicosis group (mean ± SE)

to wear the graduated elastic stockings with a pressure of 30–40 mmHg at the ankle region, and the plethysmographic examination was repeated with a transducer applied against the stocking.

There was no significant difference in either parameter between measurements without and with stockings in normal controls. In limbs with varicose veins, the $\frac{1}{2} \cdot t$ showed a significant increase when subjects were wearing the elastic stocking, indicating a reduction of venous reflux (Fig. 7). The EV showed no significant difference between measurements without and with stockings. However, when the tourniquet was applied over the stocking, no significant increase of EV was obtained. These results are similar to those after the surgery.

These findings indicate that the abnormal increase of venous pooling in the varicose group decreases after both beneficial treatments.

Chronic Venous Insufficiency

In this study, 104 limbs with varicose veins were divided into two groups—64 limbs with simple varicosis (simple varicosis group) and 40 limbs with chronic

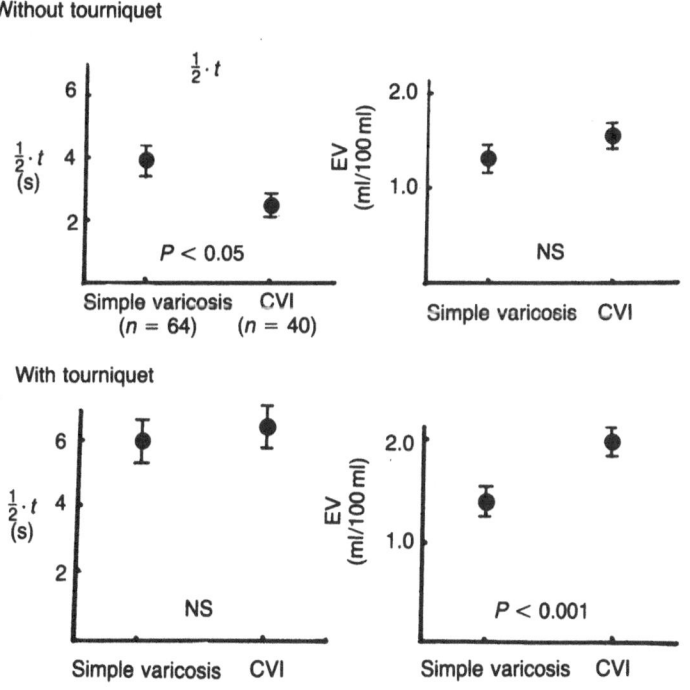

FIG. 8. Comparison between simple varicosis and chronic venous insufficiency (*CVI*) groups (mean ± SE)

venous insufficiency, such as pigmentation and/or leg ulcers (CVI group), and the pathogenesis of chronic venous insufficiency was investigated. Muscle pump plethysmography was used to compare the venous function between the two groups. The strain gauge transducers were applied at the distal part of the calf, just above the medial malleolus, because the signs of chronic venous insufficiency usually develop in this area.

The mean $\frac{1}{2} \cdot t$ in the CVI group was significantly shorter than that in the simple varicosis group (Fig. 8). With application of tourniquets, the simple varicosis and CVI groups both showed improvement in $\frac{1}{2} \cdot t$, and subsequently, the significant difference between the two groups disappeared. The CVI group showed a significantly higher value of EV with tourniquets than the simple varicosis group.

From these results, chronic venous insufficiency is characterized by a shortened $\frac{1}{2} \cdot t$, and a high value of EV with a tourniquet.

These findings indicate that severe valvular incompetence of the superficial vein system [7] and a high degree of blood pooling in the calf might be considered to be the main cause of chronic venous insufficiency. Such venous pooling leads to capillary distension, widening of the endothelial pores, and escape of large molecules such as fibrinogen. The deposition of such large molecules around the capillary forms a barrier to the passage of oxygen and other nutrients, resulting in cell death and ulceration [8,9].

References

1. Hirai M, Naiki K, Nakayama R (1991) Chronic venous insufficiency in primary varicose veins evaluated by plethysmographic technique. Angiology 42:468–472
2. Hirai M, Naiki K (1992) Hemodynamic evaluation of venous function after surgical treatment of varicose veins. Vasc Surg 26:345–350
3. Hirai M, Naiki K, Nakayama R (1992) Hemodynamic evaluation of elastic stockings for treatment of varicose veins. Int J Angiol 1:6–9
4. Sumner DS (1982) Strain-gauge plethysmography. In: Bernstein EF (ed) Noninvasive diagnostic techniques in vascular disease. CV Mosby, St Louis, pp 468–481
5. O'Donnell TF, Shepard AD (1985) Chronic venous insufficiency. In: Jarett F, Hirsch SA (eds) Vascular surgery of the lower extremity. CV Mosby, St Louis, p 206
6. Sakaguchi S, Ishitobi K, Kameda T (1972) Functional segmental plethysmography with mercury strain gauge. Angiology 23:127–135
7. Hoare MC, Nicolaides AN, Miles CR, Shull K, Jury RP, Needham T, Dudley HAF (1982) The role of primary varicose veins in venous ulceration. Surgery 93:450–453
8. Browse NL, Burnand KG (1982) The cause of venous ulceration. Lancet 31:243–245
9. Angel MF, Ramasastry SS, Schwarz WM, Basford RE, Futrell JW (1987) The causes of skin ulcerations associated with venous insufficiency: A unifying hypothesis. Plast Reconst Surg 79:289–297

Subject Index